CAMBRIDGE LIBRARY COLLECTION

Books of enduring scholarly value

Life Sciences

Until the nineteenth century, the various subjects now known as the life sciences were regarded either as arcane studies which had little impact on ordinary daily life, or as a genteel hobby for the leisured classes. The increasing academic rigour and systematisation brought to the study of botany, zoology and other disciplines, and their adoption in university curricula, are reflected in the books reissued in this series.

Chapters in My Life

Leonard Jenyns (1800–93; he changed his name late in life to benefit from a legacy), was a clergyman, and a respected naturalist and zoologist. A distinguished member of a dozen scientific societies, he was educated at Eton, and then at St John's College, Cambridge, where he graduated in 1822. During his tenure as vicar in Swaffham Bulbeck, he made important contributions to zoology, becoming one of the original members of the Zoological Society of London. In 1831, unwilling to spend years away from his parish responsibilities, he turned down the chance to travel as the naturalist on-board H.M.S. *Beagle*. Published in 1889, this is the second edition of Jenyns' autobiography, which he had first had privately printed. It starts with the major events of his life, then shares a series of scientific anecdotes, including his decision to recommend Darwin instead of himself as the naturalist for the *Beagle* voyage.

Cambridge University Press has long been a pioneer in the reissuing of out-of-print titles from its own backlist, producing digital reprints of books that are still sought after by scholars and students but could not be reprinted economically using traditional technology. The Cambridge Library Collection extends this activity to a wider range of books which are still of importance to researchers and professionals, either for the source material they contain, or as landmarks in the history of their academic discipline.

Drawing from the world-renowned collections in the Cambridge University Library, and guided by the advice of experts in each subject area, Cambridge University Press is using state-of-the-art scanning machines in its own Printing House to capture the content of each book selected for inclusion. The files are processed to give a consistently clear, crisp image, and the books finished to the high quality standard for which the Press is recognised around the world. The latest print-on-demand technology ensures that the books will remain available indefinitely, and that orders for single or multiple copies can quickly be supplied.

The Cambridge Library Collection will bring back to life books of enduring scholarly value (including out-of-copyright works originally issued by other publishers) across a wide range of disciplines in the humanities and social sciences and in science and technology.

Chapters in My Life

Leonard Jenyns

CAMBRIDGE UNIVERSITY PRESS

Cambridge, New York, Melbourne, Madrid, Cape Town,
Singapore, São Paolo, Delhi, Tokyo, Mexico City

Published in the United States of America by Cambridge University Press, New York

www.cambridge.org
Information on this title: www.cambridge.org/9781108038492

© in this compilation Cambridge University Press 2011

This edition first published 1889
This digitally printed version 2011

ISBN 978-1-108-03849-2 Paperback

This book reproduces the text of the original edition. The content and language reflect the beliefs, practices and terminology of their time, and have not been updated.

Cambridge University Press wishes to make clear that the book, unless originally published by Cambridge, is not being republished by, in association or collaboration with, or with the endorsement or approval of, the original publisher or its successors in title.

CHAPTERS

IN MY

LIFE.

CHAPTERS IN MY LIFE.

WITH APPENDIX,

CONTAINING SPECIAL NOTICES

OF

PARTICULAR INCIDENTS AND PERSONS;

ALSO,

THOUGHTS ON CERTAIN SUBJECTS.

LEONARD BLOMEFIELD
(LATE JENYNS).

For Private Circulation.

REPRINT, WITH ADDITIONS.

BATH:
1889.

PREFACE.

—o—

More copies of my "Chapters" having been asked for by friends than I had at my disposal, and some who received the Booklet having complained of its brevity, I have been induced to reprint it with additions. The new matter is mostly placed in the Appendix—as either dwelling at greater length than before on certain portions of my life, or relating to incidents in the same not mentioned at all in the former impression. I have also added thoughts on a few subjects not otherwise alluded to. This arrangement leaves the general outline as given at first—unbroken.

CONTENTS.

	PAGE
PARENTAGE, EARLY LIFE, AND CHIEF EVENTS IN AFTER YEARS	9

APPENDIX.

	PAGE
CHEMISTRY	37
DR. WOLLASTON	41
THE REV. LEONARD CHAPPELOW	46
MY LIBRARY	48
WHITE'S NATURAL HISTORY OF SELBORNE	51
WHITE AND SELBORNE	52
CHARLES DARWIN	54
COLLEGE AND THE UNIVERSITY	57
ORDINATIONS AND PARISH WORK	68
MEDICAL PROFESSION, WITH REMARKS	82
IN A THUNDERSTORM	86
ISLE OF WIGHT	90
METEOROLOGY	96
RAY COMMEMORATION AND BUCKLAND	100
THE BRITISH ASSOCIATION	103
BATH NATURAL HISTORY FIELD CLUB	105
THE WORLD SLOWLY IMPROVING	106
PARTY SPIRIT	119
RECOGNITION OF FRIENDS IN A FUTURE STATE	123
SOCIETIES I BELONG TO	127
LIST OF PUBLICATIONS	128

PARENTAGE,

EARLY LIFE, AND CHIEF EVENTS

IN AFTER YEARS.

I WAS born at 10 *p.m.* on the 25th of May, 1800, in a house in Pall Mall, London, now pulled down, but then occupied by the celebrated Dr. Heberden, M.D., Physician to the Royal Family, and accounted as one of the first in his profession at that day. My mother was his daughter, and her brother, Dr. Heberden, junr., almost of equal fame as his father in his medical career, was my uncle. Dr. Heberden, senr., had married a sister of the Rev. Francis Wollaston, Rector of Chiselhurst in Kent, an Astronomer and F.R.S., and Author of a very ingenious Map of the Stars, so constructed that an observer might determine the name of any particular star he happened to see on any day or hour. Mr. Wollaston had a large family, several of his sons being distinguished in after life in different branches of science. One, the celebrated Dr. William Wollaston, was more especially known by his researches in Chemistry and Optics. He was a great friend of Sir Humphrey Davy's, and first cousin to my mother. He was often at my father's house, and I knew him well.

The above relationships on my mother's side serve to explain that general fondness for science in all its branches which I early acquired and retained through life. My taste, however, for Natural History in particular came from a different source, as will hereafter appear.

My father, the Rev. George Leonard Jenyns, son of John Harvey Jenyns, an Alderman of Eye, might never have attained to any social standing of importance, but for the circumstance of his coming into the Bottisham Hall property in Cambridgeshire quite in early life, on the decease of Soame Jenyns—second cousin to my father, and well known in the Literature of this Country—who died in 1787, leaving no issue.—This of course established my father as one of the Gentlemen of the County.—He was a Canon of Ely, or Prebendary as it was then called, a Magistrate also, and he had a great knowledge of Farming, being Chairman of the Board of Agriculture in London. He was also Chairman of the Bedford Level Corporation and made himself useful in various ways.

The earliest occurrence in my life that I can remember is the Funeral of Lord Nelson, which took place in January, 1806, when I was between five and six years old. I well remember pictures of the funeral car being hawked about the streets, and one being brought up into the Nursery for us children to look at. I may remark that I was the youngest of three boys (as alive at that time), and had four sisters, all except one older than myself.

There is nothing further that I can distinctly call

to mind, previous to my going to School at Putney, on April 22nd, 1809, from Connaught Place, where my father had a house. The street consisted of newly-built houses, some at that time still unfinished, and it was the last street in London West of Oxford Street and the Edgware Road—all beyond being green fields. Tyburn Turnpike was still standing; a high brick wall separated the Uxbridge Road (as it was then called) from Hyde Park, with trees and underwood on the Park side of the wall where nightingales might be heard in the season.

I got the usual preliminary schooling in Latin and Greek at Putney, but did not, I imagine, show any forwardness or talent, as regards the learning of those languages, beyond others of my own standing. In another language, however—French—I had the advantage; having been taught this last by my sisters at home, and certainly knowing more of it than the other boys of my own class. A curious circumstance arose out of this. There was to be an examination in French and a prize given to the boy who passed best. None of the boys but myself having the least expectation of getting the prize, it was agreed among them that they should draw lots beforehand whose it should be, an agreement, which, without showing a good deal of self-presumption, I could not but fall in with. It turned out, however, as I had expected. I gained the prize, which of course I passed over to the boy to whom it had been allotted. He took it of me, but a short time afterwards brought it back,—saying he was ashamed to keep it. This was in 1810, the year after I first went to Putney, The prize itself was a small pocket edition of Robinson Crusoe, which I still have in my possession.

Both as a child and as a boy, I was naturally still and quiet; well satisfied to be left by myself, and when at school not joining much in the games and amusements of the other boys. This was partly due to a weakly constitution, which, in after years, much interfered with my professional work as the clergyman of a large parish, showing itself especially in frequent sick headaches that continued to trouble me at intervals till after middle life. My naturally quiet disposition showed itself in rather an unusual way, on the occasion of the family travelling from Bottisham Hall to London or on the return journey. The paternal coach not being able of course to contain the whole party, a servant and some of the youngsters were dispatched in one of the old hack-chaises of those days; in which I preferred for a time keeping quiet in the straw at the bottom of the carriage, instead of always wanting, like most other children, to stand up at the windows and look out.

I also as a boy had that fondness for order, method, and precision, which I retained through life; arranging all my things, clothes, books, etc., with great particularity;—neat and tidy in everything. I was likewise somewhat taciturn. My school-fellows nick-named me—*Methodist* and *Dummy*. This I did not like. But it was true all the same. Through life I have been a man of few words, as regards the staple conversation of ordinary society; and even to old age I have been often called a *very particular gentleman*.

Both at Putney and at Eton, to which large Public School I was removed in 1813, my thoughts were much taken up with the subject of Natural History, to which I had shewn a strong leaning, derived probably in great measure from my uncle Chappelow, who was a Naturalist and my godfather. And when at Eton, being no cricketer, nor joining—except on rare occasions—in any other sports, I preferred wandering by myself in the green lanes that skirted the Playing and Shooting Fields, looking after stag-beetles (very common there), watching birds, etc.

I may here add that I was rather fond of solitary walks through life, wrapped up at such times in *observing, admiring,* and *reflecting.* Nature, in its widest aspect, with all its charms—on a still summer's day especially—had a great attraction for me, irrespective of Natural History pursuits. And it is almost necessary to be alone on such occasions—in order to drink in, as it were, the fulness of pleasure the country yields to the observer, apart from company and conversation.

I do not think that at Eton I made such progress in the Classics as I ought to have done, except perhaps as regards Latin versifying, but the system of tutorial teaching there at that time was bad, and not calculated to advance good scholarship unless a boy had a natural turn for Greek and Latin. The best copy of verses I made at Eton was a copy consisting of sixty-six hexameters, written on the occasion of the first Arctic Voyage in the present century emanating from our government,—the subject having been selected by the Master. The ships set sail on the 4th of April, 1818. A Narrative of the Voyage, in the form of a thin octavo of about a hundred pages, was written by Lieutenant Parry, Commander of one of the ships, and was, I suppose, the earliest of all the many volumes that have since appeared in connection with North Polar Expeditions. A copy of this pamphlet, probably rare, is in the "Jenyns Library" at the Bath Literary Institution. There also, along with other manuscripts of mine, will be found the verses in question, which I remember my Tutor reading over—saying at the end, "Well, very well."

I made some pleasant acquaintances at Eton, which in one or two cases led to more permanent friendships than those of a school life. I may mention especially George Howard (Bubble Howard as he was called by the boys), afterwards Earl of Carlisle and Lord-Lieutenant of Ireland, who was not only next to me in the school, but was at the same Dame's House as myself. This led to a close intimacy between us. We were much in one another's rooms, generally breakfasted together, and kept up a correspondence for some time after my leaving Eton,—where he—two years my junior in age,—remained some time longer. Whilst at Eton he once or twice dined at my father's house in Connaught Place.

It is of interest to see how one's schoolfellows turn out in after life, sometimes so different from what one would ever have looked for. I was at Eton with the two Puseys, brothers. The younger of the two was not very much above me; he was not promising in features or general appearance. Some might have called him an insignificant-looking boy. Nor did he distinguish himself by his learning in any particular way that I remember. This unpromising boy, however, grew up to be the celebrated Dr. Pusey, founder of a party in the Church bearing his own name and known to all the religious world, upon which indeed he has left an influence not soon to be forgotten.

It is also curious how one's old schoolfellows turn up sometimes unexpectedly in after life. Henry Richards, an old Eton acquaintance of mine, was uncle to my first wife. He was a clergyman and had the living of Horfield, near Bristol, where we often visited him. On the occasion of one of our visits he had a dinner party, and he told me who were to be the guests. Amongst others he mentioned Sir John Davis, who formerly held a diplomatic post in China, and was the author of two volumes on China, which I had read and taken much interest in, but I did not remember him in any other way. When he arrived to dinner Richards introduced me to him as the Rev. Leonard Jenyns. On hearing my name he started back— saying—"I remember a Leonard Jenyns at Putney!" And immediately the past flashed back to my own recollection, with the exclamation—" I, too, remember at Putney John Davis!" This sudden recognition of each other afforded matter for much conversation afterwards;—both of us amused to think that we had run against each other in this way, not having met before since we were small boys together in a private school some five and thirty or forty years previous.

I have spoken above of my quiet habits as a boy. After acquiring a taste for Natural History I became, too, of a studious disposition. I remember the pleasure I took, while still young, in being allowed when the family went into dinner, to remain behind in the drawing-room with my books, reading and writing till they returned. This was the beginning of a studious life, kept up ever afterwards.

The seclusion in which an unmarried country clergyman passes most of his days conduces, further, to much silent thought and reflection, as appears to have been the case with myself very many years after the circumstance just mentioned. A faithful servant who had been long in my service once said to my sister, when on a visit to the Vicarage, " My Master, you know, is such a thinking gentleman."

It was early determined by myself of what profession I should be. Before nursery days were over, I remember taking two rush-bottom chairs—turning one over the other, so as to leave a space in the midst of the four legs of the chair that was uppermost,—clambering up into it—calling it my pulpit,—and then preaching, or pretending to preach, a Sermon on

Dives and Lazarus. This idea of going into the Church, as the expression is, was never changed for any other. On the very day of my attaining the age of 23, I was ordained to the Curacy of Swaffham Bulbeck in Cambridgeshire, a parish of about 700 in population—close adjoining my father's Bottisham Hall Property, and began parish work by taking two full services on the Sunday following. The Vicar of the parish (whose Curate I was to be) was a school-master, who kept a school in the neighbourhood of Wisbech, and who, I was told, had never been in the parish since the day he read himself in. And, what I should think must have been almost unprecedented in the history of our Church, he gave me the appointment without any interview, and I never saw him till shortly before his death—twenty years afterwards! In truth I was the first resident clergyman the people of the parish had ever known. I only, however, held on as Curate five years,—the Vicar then resigning, and the Bishop of Ely (Sparke) giving me the living in his place.

Considering myself then a fixture, I enlarged the Vicarage House, made a garden, planted trees and shrubs; the proximity of the church and vicarage to the family residence at Bottisham Hall made it to *me* a very desirable piece of preferment, though the value of it was not large. From the front windows there was a pretty view of the Bottisham woods and plantations not far off; while the Fens, out of view but within a walk—as also Newmarket Heath and the Devil's Ditch—afforded rich ground for Natural History pursuits.

I considered myself settled for life in respect of a residence, and a parish to have charge of, and where I preached my first sermon, I expected—and quite hoped—to preach my last. But things were ordered otherwise. I was obliged—after having held the living nearly 30 years—to resign. This was owing to the state of my first wife's health. Acting under the advice of Sir Benjamin Brodie, who attended her, I had to take her away from Cambridgeshire, the climate of which was clearly against her recovery. Of course I resigned my living under the circumstances, there being no chance of my returning into residence; to continue holding it on as a non-resident incumbent I would not allow myself to do. We removed to Ventnor in the Isle of Wight, where we spent eight months; and then to Bath, where at the expiration of ten years from the time of leaving Swaffham Bulbeck, my wife died. Bath has been my residence ever since.

I may here mention my marriages. My first wife, Jane Daubeny, whom I married in 1844, was the eldest daughter of the Rev. Andrew Edward Daubeny of Eastington House, a few miles from Cirencester on the Fairford road. He held the two small livings of Ampney St. Peter and Ampney Crucis, both in the immediate neighbourhood of his own residence. Before going to Oxford, followed by taking orders, he was in the Navy, and served under Lord Nelson in the battle of Copenhagen in 1801, when he was wounded on board the Bellona. He died in 1877 at the age of 93.

His brother, Dr. Charles Daubeny, much younger than himself, was the well-known Oxford Professor— first of Chemistry, afterwards of Botany. Dr. Daubeny's house was in the Botanic Garden and very pleasantly situated; a delightful residence in spring and summer, where my wife and self often visited him, sauntering about that well-ordered garden just as we were disposed. Dr. Daubeny had frequent parties to dinners and so forth, which led to our making the acquaintance of several other Professors and notables in the University. With the late Professors Phillips and Rolleston we were quite intimate. Professor Westwood, still living, I have known for very many years. I first met him at dinner at the house of Mr. J. F. Stephens, well known to entomologists as the author of "British Entomology," and where I remember also meeting the veteran Haworth, author of the scarce and—in its day—valuable "Lepidoptera Britannica," published so far back as 1803. Professor Earle (Anglo-Saxon) became known to me from having taken the Oriel Living of Swainswick near Bath, where I was resident at the time, I had the honour of inducting him into the Living, and he has been one of my most esteemed friends ever since.

My acquaintance with Oxford was further increased by the circumstance of the late President of Magdalene College, Dr. Bulley, marrying my first wife's sister. This led to our visiting there also, as well as at the Botanic Garden, for many successive summers, and getting acquainted with some of the College Fellows and other Members of the University. In fact, I gradually found myself almost as much at home in Oxford as I had been formerly in Cambridge.

My second marriage was in 1862, to Sarah, eldest daughter of the Rev. Robert Hawthorn, for some years Curate of Swaffham Prior in Cambridgeshire, next parish to my own,—but afterwards Vicar of Stapleford near Cambridge. Her parents were both Scotch and she was born in Scotland. Many of the above mentioned visits to Oxford were made in company with my second wife.

I think I have read that, in respect of moral and intellectual qualities, where such are developed as life advances, they are more often derived from the mother than from the father. Such I consider to have been the case with myself. My father was of a very opposite disposition to myself in many respects. He was a clergyman of the old-fashioned stamp,—had not much turn for literature,—a great farmer and fond of sporting, fond also of the society of those whose ideas and pursuits were like his own. He was the son of an Alderman of Eye. His mother was a sister of my great-uncle Chappelow, of whom more afterwards, and from whom I probably got my taste for Natural History. Perhaps it was to my father's disadvantage that while yet quite a young man, he came into possession of all the Bottisham Hall Property in Cambridgeshire, formerly belonging to Soame Jenyns, as already mentioned above. This may have much helped to divert his attention from the ways and habits suitable to a clergyman, leading him to abandon home studies for out-of-door occupations, and the management of a parish for that of a large farm.

My mother was a quite different person. A daughter of the widely-known Dr. Heberden, perhaps the most distinguished physician of the day, a scholar well versed in classical and general literature, a student of divinity also;—her mother a Wollaston—a name associated, as well then as in the succeeding generation, with scientific research,—there was much to favour the development of her mind, and habits of thought, in a right direction. She was not particulaly talented or studious; but she was deeply imbued with the same high moral and religious principles which characterized her father, and by which she was guided in the education of a large family of children. Of that family I was the youngest but one of those who attained to adult life.

I feel that I owe very much to her teaching and sound advice at all times; as I do also, though naturally to a less degree from not being so much with him, to her brother, my uncle, the second Dr. Heberden, also a physician (to the Court, attending George the 3rd till he died), and equally distinguished as his illustrious father for his extensive reading and general knowledge; the author, moreover, of several works relating to education, scripture, and the classics.

I had frequent opportunities of enjoying my uncle's society and conversation, from which I never came away without feeling myself a better man. It always seemed to me that my own mind was cast much in the same mould as his; so far as regards taste, judgment, and first principles,—the basis of all right thought and action;—without for a moment placing myself on a par with him, or measuring my own attainments by the excellency of his character. I think we held pretty much the same opinion respecting men and books, and other things. I suppose, too, besides being like my uncle in habits and similarity of thought, I must have been like him in features and outward appearance; for not only was it often remarked by members of our family how much I resembled him, but twice was I mistaken for one of his own sons;— on one occasion by a fellow traveller in a stage coach, an entire stranger, who, after looking at me awhile, said, "Pray, Sir, are you not one of Dr. Heberden's sons?"

While yet quite a young man reflecting on my father's occupations and pursuits—in connection with the clerical profession, which I had determined on for myself,—there were four things I resolved never to have anything to do with, not so much from their incompatibility with Church ministerial work, which hardly applied to all, as from my distaste to some of them, and, with respect to others, thinking they would at least draw away much of my attention from the duties incidental to the care of a large parish: also what relaxation I needed was amply afforded me in my Natural History pursuits and love for books. The four things in question were *Sporting, Farming, Politics,* and *Magisterial Business.*

The first of these calls for a little notice. I have read somewhere of the poet Gray, that when near his end he confessed he had never been upon a horse's back in his life. This is not my case. I was very fond of riding all the early part of my life, and kept a horse of my own for some time at Swaffham Bulbeck Vicarage. My confession would be that I never fired off a gun in my life, a circumstance the more remarkable from my father and both my brothers (the elder especially) being all keen sportsmen; frequent shooting parties being made up at Bottisham Hall in the season, with a fair amount of game on the estate, and keepers to look after it.

It might be thought, also, that my fondness for Ornithology would have led me occasionally to shoot such birds as I needed for the study of that branch of Natural History or to have the means of obtaining any rare species that might unexpectedly come across me. But I often went out with my elder brother, he having his gun in hand—ready for any occasion on which it might be wanted. The keeper, too, was always on the look out for any rare birds or other animals that might turn up. When the family were in London, where my father had a house,—a hamper of supplies (farm and garden produce) was sent up weekly from Bottisham Hall, in which there were generally, in addition to eatables, a few small birds, put up by the keeper for "Master Leonard." These birds, after the species had been determined and their characters noted down, were mostly skinned for preservation, the bodies being subjected to a rough dissection.

The above was in my boyish days. As years rolled on, my museum, if worthy of the name, increased in almost every department of Zoology. I very early commenced collecting insects, which grew to an ambition to get together as large and complete a collection as I could of the insects of Cambridgeshire, not confining it to butterflies and beetles, but including all the orders. The *Diptera*, very much neglected by Entomologists in general in those days, were my especial favourites; and a cabinet, which I got made for me in London to hold the entire collection, had several drawers devoted to this order of insects. On my quitting Cambridgeshire in 1849, this cabinet with its contents were passed over to the Museum of the Cambridge Philosophical Society, where I had already deposited numerous specimens of Mammals, Birds, Fishes, etc., obtained from time to time— mostly—within the County.

I was always fond of birds and of watching their ways. Cambridgeshire being for the most part open country, such species as like shelter naturally flocked to the gardens and plantations of Bottisham Hall, where I got well acquainted with their song, or other notes, nidification, etc. The smaller summer birds of passage came in plenty, and I was quite familiar with the nests and eggs of all the species.*

It was about the year 1831, or '32, that I delivered a short course of Lectures on Ornithology to the Cam-

---

* I remember the well-known Dr. Dibdin, who had the living of Exning near Newmarket, when dining at Bottisham Hall, spoke of the place as an "Oasis in the Desert."

bridge Philosophical Society, of which I had been elected a member in 1822. I had read a paper to that Society in 1825 on the Ornithology of Cambridgeshire. In the Lectures I am now alluding to, the subject of Ornithology was dealt with in a more general way. The *first* Lecture treated of the anatomy of Birds; the *second* of plumage and the structure of Feathers; the *third* was on the subject of Migration; the *fourth* on the Classification of Birds.

The above Lectures were illustrated by drawings on a large scale done for the occasion by my eldest sister, Mary;—they were the only Lectures, properly so called, I ever delivered. They were well received; but lecturing was not my *forte*, nor easy to me, as it was to Henslow, who was admirable in this way. I was not fluent in public speaking, nor could I get language ready to my tongue as wanted. Thinking and writing was more to my taste.

I have spoken of the illustrations for the above Lectures as done by my sister. My own utter inability to draw was a far greater disadvantage to me as a Naturalist, than not being a shot. It is noteworthy that Darwin, in his autobiography, laments the same deficiency in himself, saying how much the possession of this art would have assisted him in his Natural History investigations.

The absence of this gift in myself—the only member of the family who really needed it in a utilitarian point of view—is the more remarkable from its having been largely developed in *all* the other members of the family. My mother painted in oils. My father drew and painted landscapes in water-colours. Two out of four sisters did the same, or painted in other ways occasionally. My eldest sister of all was a most successful miniature painter; while both my brothers were quite first-rate artists, one of them having exhibited in the Water-colour Exhibition; the subjects on which they exercised their brush or pencil being various—old buildings, ships and sea-pieces, as well as figures of men and animals of all kinds. As a set-off, however, to my own want of skill in this way, my eldest sister was always at hand to make for me any delineations I needed, as in the instance above mentioned, as well as to assist me in many other ways in my Natural History pursuits.

I have spoken of the dispersal of my collections at the time of my ceasing to reside in Cambridgeshire, a large portion being given to the Museum of the Cambridge Philosophical Society. But some things went elsewhere. A collection of British birds' eggs, along with a small collection of the crania, etc.—of micromammalia,—a group I had much studied and written about,—I gave to the Museum at Ipswich.

A large Cabinet of British Shells, along with the Herbarium already spoken of, I took with me to Bath. The collection of shells was rather a poor one in respect of marine species, but it was very rich in land and fresh water species, containing nearly all that are known in this country;—fresh water especially, which abound in the fens, often attaining to an unusual size. I had a very successful way of procuring the smaller land species, which conceal themselves at the roots of grass, or are almost subterraneous, coming up to the surface to feed at night. It was suggested to me by the circumstance of my often finding minute shells on the under surface of large stones when they were turned over. My plan was to get large pieces of the bark of decayed trees and to lay them down in the evening in plantations and shady places, where there was a fair growth of herbage, and then examine them the next morning; when I always found many small specimens of Mollusca sticking to them beneath. These I called my *shell-traps*.

There was one group of shells—the small fresh water bivalves (*Cyclas* and *Pisidium*)—to which I paid particular attention, obtaining, by the assistance of various friends and correspondents, specimens from several other parts of England to compare with the Cambridgeshire species. On this group, after long study of their characters and habits, as observed in specimens kept alive in water for many months—thriving and breeding freely,—I wrote a monograph, with plates by Sowerby, afterwards published in the fourth volume of the "Transactions of the Cambridge Philosophical Society." A whole drawer in the cabinet above spoken of was devoted to this group, and is perhaps the most striking in the collection.

Darwin once—in a letter to me—expressed his surprise that, with all my parish work and church duties to attend to, I was able to do so much in Natural History. The secret of the matter lay simply in a well-considered arrangement of time and occupations. I was always an early riser, seldom, unless indisposed by illness, getting up later than six o'clock till past four-score years. I had also contracted a habit of turning all leisure hours and half-hours to good account, and (what I consider of much importance in all work requiring time and thought) never attempting two things at once, but, for so long as circumstances allowed, throwing my whole mind into whatever I was engaged upon as if there was nothing else to attend to; in accordance with the scriptural maxim—"Whatsoever thy hand findeth to do, do it with thy might."

No literary work can attain to excellence if—whilst engaged upon it—the mind wanders away to other subjects it wants also to consider. The train of thought required in one case interferes with the train of thought required in any other. My belief is that no two good books—or even shorter treatises—can be properly taken in hand at the same time. Finish one before you begin another is my rule. I remember once, at a party of Cambridge men in a College Combination Room,—it having been remarked that some writer had "too many irons in the fire," as the saying is,—Dr. Edward Daniel Clarke, the celebrated traveller and Cambridge Professor of Mineralogy, impetuously exclaiming—"O no, that can't be, throw in tongs, shovel and poker, all three together!" What, as a rule, follows from undertakings being carried on in such a hurried combination?—that the work, whatever it be, is not conducted with that care and concentrated attention required for doing it well;

and that some books and writings taken up at haphazard in this way, first one thing and then another,—each being commenced at the spur of the moment, but not gone on with continuously,—are often never completed. I may here summarize the publications to which my own pen has given issue. My two most important works were "The Fishes of the Voyage of the Beagle," and my own "Manual of British Vertebrate Animals." In spite of what I have just stated, I was to a certain extent compelled by circumstances to have these two works in hand together. I had made fair progress with the latter, when Darwin urgently pressed me to undertake the Description of his Fishes, which he said he could get no one else to do. Regard for my old friend, and the interest I took in all the valuable results of his celebrated voyage, induced me to comply. But the work cost me a deal of labour, and many books had to be consulted on the subject. Of Fishes I had made no previous study beyond an acquaintance with the chief kinds found on our own shores. Of exotic forms I knew nothing. Fortunately, I had the resources of the Cambridge University Library at my command, and the chief work on the Ichthyology of that day, the "Histoire des Poissons," by Cuvier and Valenciennes, was that which afforded me most—indeed nearly all—the assistance I needed. The first volume —introductory to the description of species—containing a detailed account of the structure, internal as well as external, of Fishes in general, I was obliged thoroughly to master before proceeding to the determination of those collected by Darwin, and describing such as were new. The work I had undertaken was to appear in numbers published from time to time, the chief of the new species being illustrated by engravings, and it took me a long while to complete it.

My second work, above mentioned—The "Manual of British Vertebrate Animals"—was brought to a finish, and published at Cambridge, in 1836, the Syndicate of the University Press giving me a free impression of 750 copies.

It is not for me to speak of the merits of my own production. The leading Zoologists of that day gave it a good character, and seemed highly to approve of it; but from the nature of the book—its condensed remarks on all relating to the *habits* of the animals spoken of—as distinguished from the *descriptions,*—the latter being given for the most part in much detail and with great exactness—it did not fall into the list of popular works, and attracted only Scientific Naturalists interested in the Fauna of Great Britain. I suppose between six and seven hundred copies got into circulation, but it never came to a 2nd edition; though I remember Charles Lucien Bonaparte, the great Ornithologist, whom I met at Oxford some years after its publication, and who much commended the work,—pressing me to bring out a new edition, in order to meet the advances made in the subject since the appearance of the first, leaving all the unsold copies to their fate.—I did not, however, fall in with that idea.

The above "Manual" was followed in 1846 by my

"Observations in Natural History: with an Introduction on Habits of Observing, as connected with the study of that Science: also a Calendar of Periodic Phenomena in Natural History, etc."—and after this, in 1858, by my "Observations in Meteorology."— Neither of these works, though they had a fair sale, were sufficiently popular for the general public. But I have written nothing—*otherwise than* with a view to the further extension and advantage of those Sciences which have been my study and amusement through life.

Henslow died in 1861, and the following year I brought out a "Memoir" of him, which as well as the two other books just mentioned were published by Van Voorst.

In addition to the above works, I contributed a variety of Papers and Short Articles, at different times, to the Transactions of· Scientific Bodies and to other Periodicals; the most important, perhaps, being a "Report on Zoology," read to the British Association at their Edinburgh Meeting in 1834, and published in the volume of their Proceedings for that year.

I may also mention that I wrote the article on Yarrell's British Birds in the 65th number of the "London and Westminster Review," published in March, 1840.

I have never been abroad. This may astonish many persons at the present day when all the world is moving about in every direction. But it is not everyone whose circumstances allow of his going abroad whenever he has the fancy to do so. What have been my own circumstances? During the early part of adult life I had regular parish work, which seldom permitted me to go from home, while from the small value of my preferment (for 5 years only a curacy) I had not means sufficient for much travelling. After my first marriage in 1844, the prolonged illness of my wife and the nature of her complaint only permitted short journeys to be taken—visits to friends or to some watering place that might be easily reached without much fatigue. After her death in 1860, I had more leisure, having resigned my living; but as I neared the age of three-score years and ten I felt less and less inclination to go on the Continent for the first time.—I was content to remain in England, visiting different parts of it from time to time—rich in scenery or attractions in other ways,—many of those who are in the habit of going abroad simply following the fashion,—and remaining through life— more or less ignorant of their own country.

My tours in the first instances were simply walking tours, from considerations of economy and the independence allowing one to stop where one pleases, essential for Natural History pursuits. These tours originated in the following way :— Old Mr. Wollaston, of Chiselhurst, my mother's uncle, was still living when I was a boy; and I well remember my mother, one summer season, taking myself and others of the family in an open carriage to visit him, and spend the afternoon at his Parsonage. Chiselhurst in those days was several miles away from London, and a very pretty country drive it was. In truth I was so much charmed with the woodland scenery,—in the midst of which that village was then located,—that I determined, when old enough and so circumstanced as to be able to take a tour by myself, my first should be a walking tour in Kent; and I kept to my determination.

In 1826, I visited—chiefly on foot—several parts of Kent and the Isle of Thanet, taking long walks on the South Coast, as well as inland, and ending with a second visit to Chiselhurst,—where my acquaintance with Kent had first begun so many years before.— Kent was visited again in two other years, 1827 and 1828, these tours being combined with a trip to some of the adjoining counties.

In 1830, I took a complete walking tour in Derbyshire,—visiting every place and object of any interest— exploring the celebrated Peak Cavern at Castleton, quite to the extremity, and even going down the Odin lead mines, worked ever since the days of the Saxons. This last undertaking was not without danger, there being no means of descent except that of going down a well—not in a bucket or by ladder—but by placing your feet alternately in holes, cut to receive them, on opposite sides of the well.

Other tours, in different years, were taken to many of the midland and northern counties; including a short excursion in Scotland—visiting Edinburgh and The Trossachs, Loch Catrine, Loch Lomond, and Glasgow; and a tour in 1842 to South Wales.

Many of the above tours had for their chief object Botanical and Natural History researches. In other cases they were undertaken with a view to Churches and Gothic Architecture. For this last subject I had acquired a taste in early days, when my father being a Prebendary (now called *Canon*) of Ely, and the family often there in residence,—I was much impressed with the grandeur of that magnificent cathedral, every part of which I gradually became acquainted with, and carefully studied with the help of "Rickman's Gothic Architecture."

My interest in this last subject increased each year. I subsequently, from time to time, visited most of the other cathedrals in this country, as well as a considerable number of the finest of our churches in the different styles of Gothic; besides very many small village churches, as opportunity offered, several of these containing features of great interest, and occasionally of rare occurrence.

In one instance, I so arranged my touring as to be at the morning service of six different cathedrals on six successive Sundays.

More than once, on the occasion of a northern tour, I was a visitor at Auckland Castle, the seat of the Bishop of Durham, one of whose Chaplains I had been ever since he was raised to the Episcopate—(Maltby, first Bishop of Chichester, afterwards of Durham),—and one Sunday preached in the Chapel of Auckland Castle, to a small congregation composed chiefly of visitors at the castle and servants.

After my second marriage, my wife and self

generally spent a portion of each summer in some other place than Bath. In 1865, we occupied a house at Torquay, lent us by Dr. Daubeny, Professor first of Chemistry and then of Botany at Oxford, and my first wife's uncle,—where we spent two months, enjoying the scenery of that part of Devonshire.—In 1867 we went still further west;—on to Cornwall,—visiting Penzance, the Land's End, and the Lizard. Both at Torquay and in Cornwall, I added very largely to my Herbarium of British Plants, collecting nearly all the rare species peculiar (for the most part) to the extreme western counties.

Others years we went to the Lake District, taking up our quarters on different occasions at Ambleside, Grasmere, and Keswick. In others; to Llanberis, Bettws-y-coed,—(ascended *Snowdon* from the former place in my 72nd year)—Beddgelert and Anglesea,—in N. Wales.

Then for three successive years,—1880, 1881, and 1882,—we visited the New Forest, in Hampshire, taking up our quarters at Lyndhurst or the immediate neighbourhood.

I consider *Woodland* scenery quite as charming and attractive as *Lake* scenery.

On my first coming into the Bath neighbourhood, in 1850, I occupied a house at South Stoke, near Combe Down, for two years, but had occasion to change my residence several times afterwards.

In 1852 I removed to Swainswick, where we remained for eight years, during which time I served the

Curacy of Woolley, and for a year or two that of Langridge as well. In 1860, on the death of my first wife, I changed the country for Bath itself, renting a house in Darlington Place, Bathwick; and this house was given up, in 1869, for another I bought—19, Belmont—where I am now, and where I am likely to remain for the rest of my life.

It was at the time of making this last change, that I made over, as a gift, the whole of my Natural History Library, and my Herbarium of British Plants, to the " Bath Royal Literary and Scientific Institution."

Since then, my Scientific work has been mostly in connection with the " Bath Natural History Field Club," founded by myself in 1855. That Club has now published between five and six volumes of its "Proceedings," and in them are about twenty communications from myself.

I consider, however, my work in all ways as at this time entirely finished. Natural History, combined formerly with Church and Parochial duties, has been a source of happiness to me through life. Science, books, and visiting the poor,—the three occupations I took most pleasure in—were always at hand, and each attended to in its turn. So long as I was well in health, time never hung on hand. I trust the duties of a clergyman have not been forgotten amid the attractions of other pursuits. May I be judged to have led, not otherwise than a good and useful life. Now in my 89th year, the end cannot be far off.

LEONARD BLOMEFIELD,

Bath; 1888.   (LATE JENYNS).

APPENDIX.

CHEMISTRY.

Whilst at school at Putney, an itinerant Lecturer on different branches of Physical Science, of the name of Walker, who was in the habit of visiting schools, came and delivered a Course of Lectures on Chemistry to Mr. Carmalt's boys. These lectures I attended, and I became so inoculated with the subject,—so struck with the brilliant experiments performed by the Lecturer in illustration of his subject,—that for many years afterwards that branch of science took strong hold of my mind, almost to the exclusion of Natural History. Not content with thinking and reading about it,—having purchased a few works on the subject,—I saved what money I could with a view to procuring some of the more simple apparatus necessary to enable me to try the same experiments I had seen at the lectures. I had little opportunity of doing this at a private school like Putney; but, on leaving Mr. Carmalt's school for Eton, where I went in 1813, and had a room to myself in my Dame's house, I was better circumstanced for carrying on operations of this kind. I was fortunate, too, in getting much assistance from Mr. Rogerson, an intelligent and long-established cutler at Eton, as well as from his clever foreman—well known to all the Eton boys of that day by the name of Billy. Between the two they constructed for me, or procured, much of the apparatus I wanted, while the latter co-operated with me in many ways—especially in the manufacture of coal gas, with which I lit up my room.

My father at that time was living in London, and some of my school holidays were spent in London with the family. This circumstance brought me into company with two or three distinguished men of science. One was Dr. Wollaston, the eminent chemist and natural philosopher of that day, who was first cousin to my mother, and often visited at my father's house in Connaught Place. He kindly asked me to come and visit him at his own house, in Dorset Street, when he would show me some chemical experiments. I was greatly delighted at this offer, which I accepted at once. On entering the house at the appointed time, he soon appeared, and I asked him if I was going into his laboratory;—upon which he very decidedly answered—"No, you do not go into my laboratory; I bring my laboratory to you." I heard afterwards that no one was ever admitted there. He then showed me into an adjoining room, and after a few minutes he re-appeared with a small tray in his hands, fitted up in the simplest way imaginable; a spirit lamp and blow-pipe, a few re-agents in small bottles, watch glasses, and a few plain slips of glass, like the slides of a microscope, and some glass tubes; these things made up, as far as I remember, the whole of his humble apparatus, with which he showed off some of the most striking experiments. I particularly remember one in which a precipitate was made by the help of a glass rod to write down its own name at the bottom of a watch-glass. Many of his experiments were carried out upon mere drops brought together on a slip of glass at will by a glass rod. He had a great reputation for what might be called micro-chemistry. It is recorded that on meeting a friend one day in a wide London street, he took him aside for a minute, and then drew from his pocket a tailor's thimble containing a minute galvanic battery. A few drops of acid from a vial he also had about him soon set it in action, and by it he melted a piece of very fine platinum wire, none of the passers by noticing what was going on. He was able to manipulate with very scanty material as well as scanty apparatus, and to determine the constituents of a new mineral from the smallest fragment, if from the rarity of the specimen no larger could be spared for analysis.

Another well-known chemist and mineralogist, who occasionally dined at my father's house and took notice of me, was Mr. Hatchett. His name lives in a peculiar bituminous mineral—" mineral adipocire, or *Hatchettine,"*—called after him. But the most distinguished savant of that day, in respect of rank and the position he occupied, was Sir Joseph Banks, President of the Royal Society. I think it was in 1817 that Sir George Duckett, a friend of my father's who was often at our house, and who had a large acquaintance among scientific men, took me with him to one of Sir J. Banks's Sunday Soirées, where—in Sir Joseph's large library—all the living celebrities in science in London used to congregate. I well remember seeing there, among others, Herschell (Sir

John) and Babbage, then both young men; Dr. Thomas
Young, remarkable for the extent and variety of his
knowledge, not in science only, but in languages and
Egyptian hieroglyphics; Sir Everard Home, the
celebrated surgeon and comparative anatomist. Sir
George Duckett introduced me to Sir Joseph Banks—
then aged and confined to his arm-chair by gout—as
the "Eton boy who lit his rooms with gas;" this
mode of lighting streets and buildings not having
been more than a few years previous generally adopted
in the metropolis. Mr. Brande, Professor of Chemistry
at the Royal Institution, was standing by Sir Joseph's
chair at the time,—and said, on hearing what I had
been doing,—"You must have required some apparatus
for that;"—upon which I answered I *had* the necessary
apparatus.

The year following (1818) I went to College, where
I was unable to continue my chemical operations;
and two years after my admission at St. John's,
becoming acquainted with Henslow, of the same
College, though my senior by four years,—and fond of
Natural History like myself,—I gave myself up
entirely, in companionship with him, to Entomology,
Conchology, and Botany.*

* (See my "Memoir of Professor Henslow," p. 19, and 22 seq.)

Dr. Wollaston.

Dr. William Wollaston, of whom I have spoken above, was one of the most remarkable men of his age. He was the third son of the Rev. Francis Wollaston, of Chiselhurst. The eldest son was senior wrangler, and afterwards Jacksonian Professor of Chemistry at Cambridge. Other of the sons were more or less distinguished in different ways, but none rose to the eminence attained by Dr. Wollaston. The latter was conversant with nearly all branches of physical and physiological science—attaching himself, however, more particularly to chemistry and optics. In these sciences his powers of discrimination were very great. By close analysis of the ores of platinum, in co-operation with Smithson Tennant, he was able to separate from the ore three new metals—*palladium*, *iridium*, and *rhodium*, with which his name will ever be associated.

In Optics, also, he was the first—in this country at least—to notice the dark lines in the solar spectrum, which gradually led to the development of the whole subject of spectrum analysis, the fruitful source of so many modern discoveries in optical and astronomical science.

Dr. Wollaston had also a good general knowledge of Natural History, and the structure of animals and plants. He was the first to shew me in my own microscope the circulation of fluids in the *Characeæ*. In truth, he was a man of wide information on a great variety of subjects. I heard of someone who, having asked a question of a friend, at one of Sir Joseph Banks's soirées, was told in reply—"I cannot answer your question myself, but *there* stand Young and Wollaston, and between them both they know everything.". He acted as Vice-President of the Royal Society for a short time, after the retirement of Sir Joseph Banks, and was much solicited to be President, but he declined; nor did his natural disposition and habits of life exactly fit him for that high responsible office. He was reserved, and not much of a talker; showing extreme caution as to what he said in mixed company. In answering questions, or giving his opinion on any matter referred to him, he well considered the point before replying. He was equally cautious in all his physical researches, holding back from publishing his discoveries till he felt quite sure there was no flaw, in either his experiments or his reasoning, to vitiate the result. This side of Dr. Wollaston's character was in strong contrast to the bold and speculative mind of his great friend and contemporary, Sir Humphry Davy, whose pursuits were not very dissimilar to his own. It was said in a memoir in which their characters were compared, that while Davy was ever on the search for truth, Wollaston seemed always to be avoiding error. Nevertheless, Dr. Wollaston had a highly gifted mind; he was a deep thinker. I remember Dr. Whewell remarking to me once, just after a return from London where he had conversed with Dr. Wollaston, that "it was like talking to pure intelligence."

And though reserved at times when there were many around him, in company with a friend well known to him he was open and communicative. When visiting at Bottisham Hall, he used to like a Natural History walk with me, when all his senses were on the alert to take in whatever offered itself to his notice. His remarks from time to time were pleasant and suggestive. He was struck on one occasion with the tenacity of life exhibited by a wasp that had been only half-killed, the three portions—head, thorax, and abdomen—all but severed from each other, yet each by its movements showing signs of life, and he said to me, "This seems to throw light on the doctrine of the Trinity!"

On the occasion of another walk his eyes turned suddenly to the ground, his attention being arrested by the appearance of some minute black specks, scattered here and there, which he was not satisfied with till he had taken up and examined. This proved to be only the excrement of some small caterpillars feeding upon a tree overhead, from the branches of which the atoms in question had fallen. He then turned to me and said, "Did you ever examine caterpillars' dung?" This amusing question was followed up by remarks, of which I have given an account, along with a statement of observations made by myself on the subject, in my "Observations in Natural History," p. 232.

Dr. Wollaston contributed a considerable number of papers to the "Philosophical Transactions," but I am not aware that he published anything else. Some of the above papers, less purely scientific than others, were of a very original and remarkable character, and such as to interest general readers. I would allude to two in particular: one on "Sounds inaudible to certain ears," having reference to the extreme notes of the musical scale and nothing to do with ordinary deafness. The other paper I allude to, of yet more interest, had for its object to explain why the eyes of large family portraits seem to follow you from place to place as you shift your position in the room; the explanation being that the apparent direction of the eyes is regulated by the relative position of the other features of the face. This was cleverly illustrated in the paper by drawings of faces, each having a *second different* face with the forehead and eyes omitted, so adjusted as to admit of being brought to overlap the first face at pleasure. The eyes then looked up or down, according as the overlapping face was on or off. These drawings were done by a young lady of Dr. Wollaston's acquaintance, and perhaps a connection, for I think her name was Wollaston Blake. She was more than an artist, for she sent me, if I remember right, a specimen of some uncommon plant that grew in the neighbourhood where she lived. The drawings alluded to were copied for me by a relation, and they have been often exhibited to friends at our house, alike amused and astonished to think such a thing possible. It is only lately that I parted with these drawings, giving them to a young lady artist in Bath of our acquaintance.

Dr. Wollaston was a distinguished chess-player. So was my eldest sister, and they often had fierce battles on the chess-board when the Doctor was staying at Bottisham Hall. Sometimes one and sometimes the other was the victor. On one occasion, after they had been playing a while, with long and deep consideration as to the next move, Dr. Wollaston, seeing he had no chance, suddenly put his hands under the two leaves of the chess-board, which folded in the middle, and brought them together, throwing all the pieces into confusion. Of course, this was an acknowledgment of defeat.

The Rev. Leonard Chappelow.

Mr. Chappelow was my godfather and great uncle. He was descended from the Rev. Edward Chappelow, of Diss, who married a sister of Francis Blomefield, the celebrated Historian of Norfolk, whose name and property devolved upon myself, in 1871. Mr. Chappelow was nephew to the Rev. Leonard Chappelow, formerly Professor of Arabic at the Cambridge University. His sister married John Harvey Jenyns, of Eye, in Suffolk, my grandfather on the paternal side.

My uncle himself was never married. He held the two livings of Roydon and Burston, in Norfolk, but resided chiefly in London, in a small house in Hill Street, Berkeley Square, standing back behind the other houses—entered by a curious long passage—and quite unseen from the street. In this house, which no longer exists, I often visited him as a boy—sometimes staying several days—and from him probably derived my taste for Natural History, he himself, an F.R.S. and F.L.S., being very fond of that subject, and having a considerable number of books relating to it, which I took great pleasure in reading and conning over. These books—along with all his other personal property—came into my possession at his death, and form, at the present day, the nucleus of the "Jenyns Library" at the Bath Royal Literary and Scientific Institution.

My uncle, as a Naturalist, paid but little attention to system and classification, nor troubled himself much about the distinctive characters of species:— nor was he a collector in any branch of the science.

He was a Naturalist somewhat after the way of White, of Selborne; though, living very much in London, he had not the opportunities, which White enjoyed, of getting practically acquainted with animals—or not to the same extent. He was of a sentimental turn of mind, and a great lover of poetry—both classical and modern; and his chief amusement was to note the habits and instincts of birds, etc., embodying from time to time, in blank verse, the results of his own observations, combined with what he had gleaned from books —especially books of poetry. In this way he hoped one day to get together materials for a work he long had in view; and the putting them into shape for publication largely occupied his time and thought during the latter part of his life. It was to be a long poem, bearing the title of "The Sentimental Naturalist." It was very nearly—if not quite—completed before he died, but was never published (it would seem) from lack of any publisher who would undertake it. It was much in the style of Erasmus Darwin's "Botanick Garden," and would probably have been but little read, and met with but a limited sale. On my uncle's death the Manuscript passed into my hands, and a few years back I had it carefully bound, and presented it to the Cambridge University, prefaced by a short notice of the Author, and it is now deposited among the other MSS. in their valuable Public Library.

Mr. Chappelow was born in October, 1744, and died at Brompton, in September, 1820. He was buried in Paddington Churchyard.

My Library.

My uncle Chappelow gave me when I was ten years old—not long after I first went to school—"Nicholson's British Encyclopædia," in 8 vols., 8vo, published the year before (1809). This work—containing a large number of Articles on Natural History, with figures and descriptions of animals, and which helped to foster my taste for that subject—very much interested me, and I treasured it exceedingly. It was the first book I possessed, beyond a Prayer book, and it was the foundation stone of my future library. From that time I acquired a fondness for books, and felt a great ambition to get together a good Library—more especially books on Natural History and allied subjects. For several years, of course, my means were too slender to allow of many purchases;—yet I found myself able to make a few—now and then—when at Eton,—and where I could not buy,—I indulged my taste for books so far as to go into bookseller's shops and look at the titles,—here and there taking down a volume and reading a few pages. When an undergraduate at Cambridge, where there were many second-hand booksellers and book-stalls, I made considerable additions to my small library, both by buying at the shops and also attending book-sales. At the latter I was able to get several of the older works on Natural History which only turn up occasionally, especially those of Ray and Linnæus, which I still possess. In 1820, when I had been two years at College, my uncle Chappelow died; and all his books—along with his other effects—

devolved upon me, his library containing several hundred volumes on Natural History and other subjects together. A large portion of the latter I exchanged, through a London bookseller, for other works on Natural History, retaining, however, some, and later in life presenting others—chiefly between fifty and sixty volumes of Italian Literature—to the Cambridge University Library, which did not possess any of them. My own Library was now getting by degrees well stocked in all departments of Natural Science; and as I became older, and had more money at my command, I made more frequent additions, several London booksellers being in the habit of sending me their catalogues. As a member, also, of several of the London Natural History Societies, as well as an old life-member of the British Association, I was entitled to all the publications these Societies issue. I may add that, on resigning my living of Swaffham Bulbeck in 1853, my Parishioners presented me with 49 volumes of Divinity allowed to be selected by myself,—which, together with the books I had before of a miscellaneous character—not scientific— made up quite another small Library, supplemental to the Natural History Library.

These contributions, coming in through different channels, so filled all the space in my house I could allot to books, that on changing my residence in Bath, from Darlington Place to Belmont, in 1869, I determined to present the whole of the scientific portion of my Library—mostly works on Natural History—to the Bath Literary Institution, *on certain conditions;*

the chief one being that the books should not be mixed up with the existing Library of the Institution, but that they should have a distinct room appropriated to them. In the same room I said I would place as a gift my entire Herbarium of British Plants, consisting of forty folio volumes, besides others in quarto.

This offer having been accepted, a room was built for the purpose, though some years elapsed before the room and its fittings up were all completed, so as to allow of the books being properly arranged on the shelves. In due time this was effected; and it was next agreed by the Committee that the room should bear the name of the "JENYNS LIBRARY," which words were painted accordingly on the outside of the door leading to it.

The number of volumes in the Library at the time of presentation was 1,200. It has since been increased to more than 2,000. This number, added to that of my own private Library at home, would stand at about 3,050 volumes.

WHITE'S NATURAL HISTORY OF SELBORNE.

Soon after going to Eton, I made the acquaintance of Lord Brecknock (son of the Marquis of Camden, of the Wilderness, Kent), who was at Eton the same time as myself, and who boarded at the same Dame's house. He had a good Library for a schoolboy; and among his books was a copy of "White's Selborne" (Markwick's 8vo edition in 2 vols.), a book I had never seen before nor even heard of,—White's name being but little known in those days. His Lordship was kind enough to lend me the book, which I read with avidity, it being so entirely in keeping with my own fondness for Natural History, and out-of-door observations on the habits of birds and other animals. Nor was I satisfied with reading it once or twice; but under the apprehension that I might never see the book again, I copied out nearly the whole of it, a few chapters only being omitted, which were of less interest in a Natural History point of view.* This MS. remained by me for a number of years, and I so often had recourse to it, previous to my being possessed of the book itself, that I nearly got it by heart. I little thought then of being the owner, some future day, of the numerous editions that would appear from time to time of that popular work,—still less of myself being the Editor of one of them.

* It is remarkable that the same thing should have been done by another person, as mentioned by Sir John Lubbock in his interesting little book, "The Pleasures of Life," p. 60.

WHITE AND SELBORNE.

In June, 1874, I went by invitation to visit Mr. Bell of Selborne, then residing in the same house, "The Wakes," in which had lived formerly Gilbert White. Mr. Bell was possessed of many interesting relics of that far-famed Naturalist. I never had a more pleasant time than the three days I spent at that classical spot and its surroundings, which I had never seen before. I was shown the room in which White wrote his Natural History of Selborne, and the room in which he died. I sat in his own arm-chair, his own old barometer hanging on the wall close by, put on his own spectacles (glasses, however, broken), and read one of his own MS. Sermons preached at Selborne Church. I visited his grave in the churchyard, where stands the venerable large yew tree he speaks of in his Natural History, close adjoining the "Plaistor"; the Church, too, was of great interest, with "Selborne Hanger" not far behind, and forming a background to his garden. The whole together was a most lovely picture, all the more attractive just then from the season and fine weather that prevailed at the time.

There are a few old Family Pictures at the Wakes, but no portrait of Gilbert White himself exists. I think Mr. Bell, in his Life of White, mentions his being very much opposed to having his picture taken. He died in 1793, and it is not likely that there be any man now living who ever saw him. I was told, however, by the late Mr. Philip Duncan, of New College, Oxford, who spent all his latter days in Bath, and whom

I frequently visited and conversed with, that he remembered having himself once seen White and been in his company. Mr. Duncan died in 1865 at the age of 93. On asking him what sort of a man White was —as to height, figure, and general appearance—he answered, to my great amusement, " O, much such as you are."

In White's garden there were some remarkably fine Lupines, which the gardener told me grew luxuriously in that soil. Giving me some seed, I brought it home for my wife, skilful in raising plants, and from that seed we have had a succession of fine plants in our garden up to the present time. We call them the Selborne Lupines.

Charles Darwin.

I became acquainted with Charles Darwin, the celebrated Naturalist, while he was yet an undergraduate at Christ's College, Cambridge. He came up to the University in 1828, and took his Bachelor's degree in 1832. Few, I imagine, now living, except his own relatives, could have known him longer than myself. He was my junior at College by ten years; but from the similarity of our pursuits, we soon became intimate after the first introduction. He was at that time a most zealous Entomologist, and attended but little—so far as I remember—to any other branch of Natural History. He occasionally visited me at my Vicarage, at Swaffham Bulbeck, and we made Entomological excursions together, sometimes in the Fens—that rich district yielding so many rare species of insects and plants—at other times in the woods and plantations of Bottisham Hall. He mostly used a sweeping net, with which he made a number of successful captures I had never made myself, though a constant resident in the neighbourhood.

He was always one of the foremost to join Henslow's "Herborizing Excursions," undertaken mainly for the benefit of the men who attended his Botanical Lectures, but open to all Naturalists who liked to go—as many did—for the opportunity thus afforded them of following up their particular pursuits in concert with others of the same taste. He was also one of those who were most constantly present at Henslow's Weekly Scientific Soirées, his own recollections of which will be found in my "Memoir of Professor Henslow."*

It was soon after taking his degree that Darwin had the appointment of Naturalist in the "Beagle," the vessel in which the late Capt. Fitzroy made his surveying voyage—extending over five years. The appointment arose in this way. Dean Peacock, at that time Fellow of Trinity College, was intimate with Captain Fitzroy, and was applied to by the latter, as to whether he could not find some one among the Cambridge men, who would be fit and willing to accompany him in his voyage in the capacity of Naturalist. Peacock immediately thought of Henslow and myself. Henslow, however, being a married man with a family, was not disposed to go under his then circumstances,—(though earlier in life no doubt he would have caught at such an offer gladly)—and he tried to persuade me to go instead. I hesitated; and, after a full day taken to consider my decision, I also declined, as well on account of my being engaged in parish work—as Vicar of Swaffham Bulbeck—which I did not think it quite right to quit for a purpose of that kind, as on account of my judging that I was not exactly the right person, either in point of health or other qualifications,—to offer myself for the situation. We then agreed (Henslow and self) that Darwin, in all respects, would be a fit man to go, and on his assenting, his name was at once sent up to Capt. Fitzroy, and the appointment was confirmed. The result showed that no better man could have been chosen for the purpose. It was during this

* *Memoir*, pp. 52, 53.

voyage that Darwin collected the materials, which—along with his own observations and thoughts on all he had seen and noted down—formed the basis of those celebrated works, which from time to time appeared in after years, testifying to the grasp his master mind had taken of Natural History in all its departments, and quite revolutionizing the whole science of Biology as then conceived. After his return, J saw him only at intervals: from the state of his health he lived a good deal in seclusion. But a correspondence was kept up for some time, and his letters to me—all preserved—are bound up, along with those of other Naturalists, in four volumes,—now in the "Jenyns Library," at the Bath Royal Literary and Scientific Institution.

COLLEGE AND THE UNIVERSITY.

From Eton I went to St. John's College, Cambridge, in 1818. There I went through the ordinary courses of a University education, but owing to my weak health—sick head-aches, before alluded to—I could not apply myself sufficiently to Mathematics to be able to take any other than an ordinary degree. I obtained, however, two prizes at the College examinations. There was no Classical Tripos in those days.

I did not make many new acquaintances at College in the first instance, except that of Henslow, of whom more presently. There were a few men, however, who had been my school-fellows either at Putney or Eton. One of these, who had been on friendly terms with me at Mr. Carmalt's school, at Putney, was Henry Malden, a studious boy, who rose in after life to be Professor of Greek in University College, London. We were of the same year at Cambridge, he being of Trinity and I of St. John's. While both yet freshmen, and but lately come, he accidentally heard of my being at St. John's, but would not come and look me up immediately, waiting to see whether I was in the first class at the examinations of freshmen at the end of the first term. Finding it so, he called, and we often associated together from that time. An amusing incident; it being no mark of very high scholarship to pass the slight examination required at that early stage of the University course.

Henslow was a Johnian like myself, and had taken his Bachelor's degree in 1818, the January previous to my first coming up in October that same year. My first acquaintance with him arose from the circumstance of his being intimate with another graduate of St. John's, who was private tutor to my cousin, George Heberden, of the same year as myself. The latter happened to mention to his tutor the circumstance of my fondness for Natural History, and the tutor mentioned it to Henslow, who—being a naturalist himself—very soon came and called on me in consequence.

Henslow had studied British Zoology under Dr. Leach, of the British Museum. Our acquaintance soon strengthened; and he was my associate ever afterwards in Natural History pursuits, till the time of his death, in 1861. He became my brother-in-law in 1823. After his death I wrote and published a "Memoir" of him, in which will be found much relating to our first acquaintance with each other, as also to my own early Natural History pursuits. These pursuits, as far as regarded the animal world, were taken up at a very early period of life—indeed, when I was scarcely more than nine years of age. But I never took up Botany till after Henslow's appointment to the Cambridge Botanical Chair, or perhaps two or three years previous, as I have specimens in my Herbarium of Plants, gathered and dried by myself, when I was quite a beginner, in 1818. We worked at the subject together, and often went on botanizing expeditions. In after life I made several tours, with a view to collecting the plants in other parts of England, and my Herbarium gradually grew, until at this present time (1888) it consists of more than forty large folio volumes of phanerogamous plants, besides several smaller volumes of mosses, sea-weeds, and fresh-water algæ. This valuable collection, embracing many very rare species, has been deposited in the "Jenyns Library," mentioned above.

My love for science having shown itself many years before I went up to the University, it was natural that whilst at college I should take an interest in those Professorial Lectures which treated of science in several of its branches, and were open to all members of the University indiscriminately. Those I selected for myself were the lectures by Professor Cumming, on Chemistry ; those of Professor Farish, on the Application of Chemistry to Arts and Manufactures ; those of Doctor Edward Daniel Clarke, on Mineralogy ; and those on Anatomy, by Professor W. Clark. The two last professors, from being of the same name, had, for distinction, the *sobriquets* applied to them of Stone Clarke and Bone Clark. I also attended a course on Geology by Professor Sedgwick.

Cumming's lectures on Chemistry interested me extremely, that being the subject I had taken to so early in life, after hearing Walker's lectures at Mr. Carmalt's school, at Putney,* and which afterwards, at Eton, I studied practically. Henslow, whom I had only just begun to know, was at that time helping Cumming as an assistant in his experiments. I also remember Sedgwick making his appearance in the room, just at the conclusion of one of Cumming's

* See Appendix, p. 37.

lectures, and this was the first time of my seeing him, when he must have been about thirty years of age. From his intimacy with Henslow, I soon got to know him well.

Professor Farish's lectures were on Chemistry as applied to Arts and Manufactures. The lecturer, however, took rather a broad view of his subject, which he is allowed to do by the conditions of the Professorship, now termed that of "Natural Experimental Philosophy." Farish brought his lectures to bear upon Coal and Mining, Pottery, and the several Fisheries, etc., many of them exemplified by machinery seen at work, which added much to their interest, and made them very popular.

I took copious notes whilst attending the lectures of Farish and Cumming, which may be found in interleaved copies of their respective syllabuses—now in the Jenyns Library.

I took the more interest in Professor Clark's Anatomical Lectures from the close alliance between Human Anatomy, to which his professorship had exclusive reference, and Comparative Anatomy—some knowledge of the latter being indispensable to the Naturalist. In instances, when the lecturer failed in getting a human subject for dissection, he would illustrate his subject by reference to the structure of animals, in addition to dried human preparations in the Anatomical Museum.

On occasions also he took us to Addenbrooke's Hospital to witness operations in connection with the particular subject on which he had been lecturing. A very few of these operations I witnessed; but I was obliged to leave the room in one instance, not being very well at the time, and beginning to feel faint.

Perhaps the most popular in those days of the public lectures I attended, as judged by the numbers that crowded in to hear them, were those of Daniel Edward Clarke, on Mineralogy. This was not so much due to the subject, as to the fame and celebrity of the lecturer. Dr. Clarke had been a great traveller; he had travelled over nearly the whole of Europe, as well as various parts of Asia. Wherever he went he had made extensive collections of minerals, medals, and valuable antiquities, as well as other things. All these were brought to bear upon his lectures, which were further set off by an amount of eloquence and declamation, such as is rarely heard in lecture-rooms : on some occasions he would seem to be almost "caught up to the third heaven." He died the same year I took my degree (1822), being succeeded in the Professorship by Henslow, whose lectures, of a purely scientific character, and delivered in plain, unimpassioned language, were as opposite to those of Clarke's as the two poles to each other. Henslow's lectures, all the same, were far the more useful of the two to a student really in earnest as to learning and taking up the science of Mineralogy.

Sedgwick had only been appointed to his Professorship in 1818, and the course of lectures on Geology which I attended, I think, must have been the first he ever delivered. He had not, before his appointment, given much attention to Geology, but he soon rose to be a master on the subject.

Henslow had a large acquaintance with the University men of his own standing, as well as of many senior to himself; and, through him, I, too, soon had a circle of acquaintances, including some of the most distinguished men in Cambridge at that time, besides the Professors above spoken of; Whewell, Julius Hare— of whom Bunsen, after his visit to Hurstmonceaux Rectory, said, "he thought he must be the most learned man of the age"—Thirlwall, Lodge, of Trinity (afterwards of Magdalene), Librarian of the University Library; Power, of Clare, who succeeded Lodge in the Librarianship; Romilly, the University Registrar; Calvert and Ramsay, both of Jesus, and both intimate friends of Henslow's, besides others I cannot call to my recollection.

Some of the above—I remember especially Whewell, Hare, and Thirlwall—paid a visit one day to my church and to my vicarage. Thirlwall distinguished himself on this occasion by taking a good run and jumping over a sunk fence that separated my garden from the meadow in front of the vicarage. I never thought at that time of the jumper rising afterwards to be Bishop of St. David's, still less of his ending his days in Bath, to which place I was myself to be removed in a few years.

I called on Bishop Thirlwall once or twice after his coming to Bath, but he was too ill to see me. On a later occasion I was asked to luncheon; but on the very morning of the day for which I was invited, he had a sudden stroke which deprived him of his eyesight, and shut him off from all further communication with the outer world. He continued to live, however, for some little while after.

The object of the party from Cambridge, more particularly, was to inspect my church of Swaffham Bulbeck. Gothic architecture was a subject to which several members of the University were giving their attention at that time, particularly Whewell. The latter was intimate with Rickman, the author of "Architecture in England," a work that greatly led to the restoration of the Gothic style in this country, as also to at least two works on the churches of Cambridgeshire, Rickman and Whewell together took a tour on the Continent to visit the churches of Normandy and other parts, during which they met with an amusing adventure at one place, where they were taken up by the National Guard, and carried to the Mayor's house, being considered as dangerous people who were drawing all the churches. They were released, however, on Rickman producing his card.*

I remember Rickman once dining in the hall of Trinity College, when I was myself one of the guests and sat next him. He was a little fat man and a Quaker, good humoured, and up to any amount of chat. I may be mistaken, but I am under a strong impression of his having told me in conversation that he had had five wives.

From my acquaintance with Whewell, Sedgwick, and other Trinity men, I was not unfrequently asked to

* The above incident is mentioned in his "Life" by Mrs. Stair Douglas, p. 147.

dine in the College Hall, and it was amusing on occasions to hear fierce arguments between some of the top men of that college. I remember Whewell and Sedgwick, at one time, getting very warm in a dispute about the Americans, though I forget the exact matter on which they were wrangling. I considered it, when the affair was over, as a drawn battle. On another occasion, Whewell and Clark (Professor of Anatomy) had a skirmish, in which, to my thinking, Clark, who was a very close and exact reasoner, had the best of it. Whewell was rather more loose in his language, very powerful, but sharp and rapid, as if wanting to flood you with words. I remember once Whewell being set down by Miller (Professor of Mineralogy), who convicted him of a mistake; the more remarkable, as Whewell himself had held that Professorship previous to Miller, and I think it was on some question of crystallography that he was wrong. Miller was a quiet little man of few words, but very clear-headed, who always spoke to the point, and thoroughly knew what he was saying.

The fact is, Whewell grasped at too much. His aim was to make himself master of the whole round of the sciences as well as of literature, and to stand first in everything; in short, to pass as another "Admirable Crichton." And no doubt in some things he stood very high; but those who for the most part restricted themselves to fewer subjects of research—if he entered upon their particular field—would sometimes trip him up. None of Sydney Smith's witty sayings were more happily conceived, or more true, than what he said of

Whewell: "Science was his *fort*; omniscience his *foible*."

I was once present at the supper given by the Master and Fellows of Trinity College, on New Year's Eve, to which many guests are invited from other Colleges, etc. The supper, I think, was in the Combination Room, and when ended, mulled wine was passed round, which—with conversation—occupied the time till midnight. On the college clock striking twelve, the doors of the room were suddenly thrown open, and the chief Butler, in proper dress, advancing with a full glass of wine in his hand, announced to the party that the old year had now expired, and the New Year come in, and he drunk to the toast of Trinity College, upon which all the guests rose—each with his glass in his hand—and did the same; a few hurrahs followed, the party immediately afterwards breaking up and going to bed.

Whether this custom is still kept up, which I think it probably is, I am not able to say.

The same year I took my degree (1822) I was elected a Fellow of the Linnean Society, of London, and also a Fellow of the Cambridge Philosophical Society. The latter Society had been established only three years previous; I have spoken of the circumstances connected with its origin in my Memoir of Henslow.* With it there was also originated a Museum, which was mainly got together, in the first instance, by Henslow and myself; the only two Naturalists in Cambridge at that time known to me, except, perhaps, Garnons, of Sydney, who collected insects. I obtained

* Page 17.

for the Museum a considerable number of the Cambridgeshire birds and other animals, but it has been greatly increased since by the addition of several entire collections of deceased Naturalists. There was no Professorship of Zoology till 1866, but it had been talked of a long while before, and I remember Sedgwick, accompanied by some other member of the University, coming over to my Vicarage to ask if I would stand for it, there being no other person they could think of, as Henslow was already Professor of Botany. I declined —on the same grounds as those on which I had declined accompanying Capt. Fitzroy as Naturalist in the voyage of the "Beagle;" viz., that I should have to give up so much of my professional work as a clergyman, if not to give up residence in my parish altogether, which I did not consider right. If the two offices were held together, the duties of neither, in my opinion, would be properly discharged. The University, I feel sure, was not a loser by my holding back from acceptance of the offer so liberally made to me. No more satisfactory appointment could have been made than that of Professor Alfred Newton, of Magdalene College, who has now held the Professorship for more than twenty years, and shown himself, in every respect, "the right man in the right place."

Before ending my University reminiscences, I cannot forbear mentioning one little incident in my College life, though it tells against me. It was the practice of the Dean of the College Chapel to call upon the scholars in turn to read the appointed lessons. One day the proper party was not there, and I, who would have sat next to him, felt called upon to take his place. The lesson happened to be a chapter in Ruth. This book, which consists of but four chapters, is not easily turned to in a moment, unless the reader is well up in the arrangement of the several books of the Old Testament, which I was not at that early period of my life. I looked for it, but could not find it. I got flurried, and after vainly turning the leaves backwards and forwards without success, sat down and hid my face in my hands, not daring to look up till I heard the voice of the Dean reading the lesson in my stead. At very few times in after life did I ever feel so thoroughly ashamed of myself as on that occasion.

ORDINATIONS AND PARISH WORK.

I was ordained Deacon in 1823, by Pelham, the then Bishop of Exeter, in Old Marylebone Church, in London. Priests' orders were conferred on me the year after, strange to say, in the Chapel of Christ's College, Cambridge. It was, perhaps, a solitary instance of an ordination taking place in a College Chapel; but in those days the Heads of Houses were not unfrequently Bishops, and the Head of Christ's at that time was Kaye, Bishop of Lincoln, to whom I had letters dimissory from the Bishop of Ely, the latter having no ordination just at that time.

After my first ordination I entered upon parish work immediately. I have stated (p. 17) that there had been no resident clergyman at Swaffham Bulbeck before myself. In truth, a resident clergyman in those days was the exception, and not the rule, in most of the villages within easy ride from Cambridge; the Sunday duty being often taken by Fellows of Colleges, and the parish left to itself the rest of the week. This led to great neglect. In my own parish, consisting chiefly of farmers and day-labourers, I found things in a deplorable state. Religion seemed to be more a matter of form than anything else. There was not a very bad congregation on Sundays, but scarce any show of reverence on the part of those who attended the services. At other times there was a profaning of the whole building, even on the part of the churchwardens themselves. There being no regular vestry, parish meetings were held mostly in the chancel, the communion table being often used on which to lay the books and settle the accounts, while some of the farmers lolled upon the rails, it might be, with their hats on their heads. In the churchyard the most abominable practices were carried on. In one instance, I found it actually used as a knacker's yard. A dead cow had been suspended by the heels against the west tower of the church, and was in process of being cut up.

To pass from the church to the school, the latter, if school it might be called, was in the most unsatisfactory state. There were only a few quite young children, the schoolmaster's wife teaching and looking after them; while the master himself, a careless and unprincipled man, was idling about the streets, doing little or nothing—perhaps in the public-house.

This school gave me more trouble than anyone can imagine. It was a charity school, founded by a benevolent lady, Mrs. Towers, a generation or two back, who left a sum of money to build a house for the Master, a school-room forming part of it. The vicar or curate of those days (I forget which), Mr. Hill, was appointed to carry out the directions in her will. As things were when I first came into the parish, the inefficient Master above spoken of had possession of the house and refused to give it up. Nothing was to be done but to enquire after the title-deeds, as to which no one living in the parish knew anything about. I had to ferret them out; and it took me months and months—if not more than a year—before I could discover where they had been deposited. I had to carry on a voluminous and most troublesome correspondence with various parties, in order to get at the history of Mr. Hill, before mentioned, to ascertain where he died, and whether he had left any existing record as to the school at Swaffham Bulbeck. Eventually, I found out that, having been formerly a fellow of Emmanuel College, Cambridge, he had taken the college living of Thurcaston, in Leicestershire, where he died. I then had to discover his will, from which, in connexion with an entry in the parish register of Thurcaston, it was ascertained that, on a certain day, he went to Swaffham Bulbeck to set up the school founded by Mrs. Towers, and that the title-deeds of the house, etc., were deposited in the treasury of Emmanuel College. At last, I thought for a moment that my task was at an end, and all difficulties removed. But not so. I went, of course, immediately, my brother—who was a barrister—accompanying me, to the Master of Emmanuel, to solicit leave to search the College Treasury for the documents so much wanted. Old Dr. Cory, the head of the College at that time, demurred at having all the contents of the muniment-room turned over for such a purpose. He consented, however, at last, on our telling him that the legal authorities declared it was of great importance, and must be done—the only condition being that the Senior Fellow of the College should accompany us. Of course, there was no objection to this. Indeed, the more the number joining in the search, the better the chance of finding what we wanted. So three of us set to work, and looked closely into every paper and document we could put our hands on, but without any success. Nothing of the kind could be found. On reporting this to our legal adviser, he was not satisfied, and said we must look again. To get leave from the Master to go over all this ground a second time was no easy matter. He made objections that caused much delay. We won him over, however, at last, and it ended in our ransacking the Treasury—not a second only, but, if I am not greatly mistaken, a third time—turning over every scrap of parchment and paper that could be found, until we all felt sure and positive that the title-deeds we were in quest of were not there, and that if they had been deposited there in the first instance they had been abstracted. I had reason to believe that this was the case and the solution of the whole question. From certain circumstances, there were strong suspicions that the Master of the School, at some time or other, and on some pretence, had got the deed into his own hands and destroyed it.

Such being assumed, there was nothing further to be done except applying to a Court of Equity. I was determined not to let the matter rest, and I had actually written a letter to Lord Brougham—Lord Chancellor at that time—and all but posted it; when suddenly there came a messenger from the village to inform me that the school-master—who from his habits was often out of pocket, and glad to do a day's work if he could get it—had been digging in a *clunch* (or chalk) pit, just outside the village, when a mass of chalk had fallen down and killed him. I hastened to his house, where he had been just brought in, and laid upon the bed—still breathing, but utterly unconscious. I remained with him till he died—only a very few hours after.

Melancholy as this event was, I hoped that it would put an end to all my troubles. But no, not yet. After her husband's funeral, the widow persisted in retaining the house, and for a considerable time after would not give it up. In the end, however, over-awed by a lawyer and his officials, she consented to retire from the struggle, and passed it over to the Vicar and Churchwardens, the Trustees appointed by the foundress in her will.

I appointed a new schoolmaster as soon as circumstances allowed, and made arrangements for the supply of what had been so long a great want in the parish. Reserving the whole house as a residence for the master,—with the aid of a grant from the authorities in London—I caused to be erected a new school-room close adjoining the house, and gradually got together a fair school, increasing in numbers yearly.—I established also a Sunday School, and this was followed after a time by the usual village clubs for clothing, coals, etc.—All these things, accompanied by the habit of frequently visiting the poor in their own houses, and performing all the Church services as earnestly as I could, brought about by degrees much improvement in the parish and in the conduct of the parishioners. Though averse to self-laudation, I might,—in order to shew that my labours had not been fruitless—refer to the testimony of the Bishop himself (at that time Turton), who, shortly before I was obliged to resign the living, remarked to the late Professor Selwyn that he considered Swaffham Bulbeck to be one of the best-regulated parishes in the Diocese.

On first commencing the care of a large parish, immediately after ordination, I felt very much the want of other clergy to consult as to the right way of doing things. I was young in the profession, and of course wholly inexperienced; and had to feel my own way, and judge for myself, as best I could in all matters. Thus—as an instance—when first called upon to go and read prayers to a sick parishioner, I said to myself—" Am I expected to take the surplice with me, or not?" My indecision and ignorance arose from there being so few resident clergy in the neighbourhood. Within a walk from the Vicarage, there was but one I was acquainted with,—a pleasant companion at times, but not a man whose manners and principles I approved of,—or of whom on any occasion I would have asked advice: his ways and habits were so essentially different from my own there was nothing to be learnt from his book. Some years after, he was obliged to leave the neighbourhood under very discreditable circumstances.

Perhaps it was better in the end that I should have been left to myself in this manner. It accustomed me to reflecting more thoroughly on what I had to do, and made me more self-reliant; weighing reasons for this or that course of action, and so gradually arriving at " a right judgment in all things," especially in cases in which there was any thought of departing from a beaten track.

A clergyman, on first coming into a parish, where things have been left very much to themselves, generally finds something in the Church services he would wish to see different; and such is what I particularly allude to above. In such cases, however desirable—or even necessary it may be—to make changes, it should be remembered that, as a rule, congregations—especially such as consist for the most part of ill-educated farmers and villagers—are naturally averse to changes, and that it is necessary to proceed very cautiously in the matter, in order not to disturb the unanimity of the parish.

My own practice on these occasions was, in the first instance, to talk the matter over with the Churchwardens, and, if possible, bring them round to the same opinion as my own. I would then mention it to a few other of the more respectable parishioners; and if it was a change of much importance, proceed next to making it the subject of a sermon, which brought it under the consideration of all the church-goers.

It was in this way that I effected a change in the time for baptisms. There had been no fixt time for performing them, as regards day or hour, which was often a great inconvenience. I wished to bring them more into accordance with the Rubric by arranging that they should take place after the Second Lesson in the afternoon service, on a particular Sunday in the month, unless there were special reasons for doing otherwise. There was some opposition at first to this innovation, but I waited my time;—waited, indeed, until some of the parishioners asked me—as if impatient to see the change carried out—" Sir, when are you going to begin having baptisms during the service?" I thought, *then* was the time to begin, and, accordingly, appointed a certain Sunday for the purpose. When the day came round, and I entered the church for the afternoon service, I found a full, almost a crowded congregation already assembled, attentive and well-behaved in every respect. The second lesson being over, I proceeded to the Font to baptise the children that had been brought, and then read the Baptismal Service as impressively as I could, and so as that all present might hear it. The result was most successful. All opposition had ceased; and the Sunday afternoons for Baptisms were ever afterwards those on which I could best reckon on having a full church and an attentive congregation.

My way of dealing with any show of ill-behaviour in church, on the part of the young men and boys (not at school), who generally sat together on short benches under the north wall, was this :—The moment I noticed any laughing or talking immediately to make a dead stop, whether reading prayers or preaching, and fix my eyes on the offender. The rest of the congregation, noticing this, immediately turned their eyes also towards the same party, who was thus shamed to a sense of what he was about, without a word spoken in reproof by anyone.

In one instance, I took a more decided step in order to enforce proper behaviour in the house of God. It was not during any of the Sunday services, but on the occasion of a marriage—when, in a village church, there are always a few idle lookers-on, more or less disposed to joke and laugh at the couple about to be united. In this instance, however, it was the bride and bridegroom themselves who misbehaved. I saw signs of it on their first coming up to the rails of the Communion Table, and cautioned them on the subject. But the caution was unheeded, and very soon after commencing the service I had to speak to them again. They still paid no attention, and I spoke to them a third time, telling them that if I saw any further signs of such levity and inattention to what they were about, I would instantly stop the service. I went on for a time, but soon saw occasion to carry out the threat, and was as good as my word. I shut the book, and—opening the gate in the rails—instantly walked straight back to the vestry, saying nothing, and there I waited for a few minutes, leaving the couple and their friends to reflect on their worse than folly and its consequences. In a short time after, the clerk came down to the vestry on behalf of the couple, saying they were extremely sorry for what had happened, and begged I would return and finish the service. I did so immediately, and all went on as it should do to the end.

The above incident—if it were somewhat of a bold step—took good effect in its consequences. I never after, from that time, had reason to complain of any misbehaviour on such occasions.

And as time went on, things improved greatly, both in the church and in the parish generally. There had always been a fair congregation, but it was much increased latterly. There was no regular meeting-house in the parish. There were some Dissenters, of course; it could hardly be otherwise expected in a population of between seven and eight hundred. But there were not very many, and some of them were among the steadiest and best-behaved of my parishioners. They sent their children to both work-day and Sunday school; and occasionally appeared at my church, where—once or twice—I even saw the Dissenting minister himself, who at times held forth in one of the cottages to the few that came to listen to him.

There were scarcely any very bad characters in the place, though reformation of manners was much needed in the case of some of the men—young men chiefly. There were only three public-houses in the parish when I first came to it, and I left the same three when I came away without any addition to the number. This was after an incumbency of thirty years. It had been attempted to set up others, but I resisted; and the farmers, who were guided by my example, withheld their consent also. In the parish nearest to my own— (where, however, there was certainly a larger population)—the number of public-houses was twenty or more.

I consider the circumstance just mentioned as showing the moral influence a clergyman may acquire over his parishioners, if he reside among them a sufficient term of years. It is seldom that any lasting good can be effected in a parish by those who, from some cause of dissatisfaction or other, seek to change their living or curacy, as the case may be—perhaps not once only, but twice or oftener.

I must say I had no fault to find with my parishioners in any way. They were, as a rule, civil and attentive to all I said and did, or tried to do, for their moral and religious improvement. For the most part, they fell in with such arrangements about things as I had to propose, and were generally ready to adopt my suggestions in respect of what I wanted new or different from what had been before. I never made or had, that I am aware, a single enemy in the place. The parish bore a good character from outsiders for peace and unanimity amongst its inhabitants.

There were great lamentations at my going, rendered necessary (as before stated, p. 18) by my first wife's bad health. And I may mention, in proof of the regard my parishioners had for me, their presentation of forty-nine handsomely-bound volumes of Divinity— leaving it to myself to choose the particular books— when I came away.

Very soon after coming to Bath—or, I should say, into the neighbourhood—and wishing still to have some professional duty, I took the small curacy of Woolley, annexed to Bathwick; the latter being a large parish—not in Bath itself, but on the other side of the river, belonging to the Duke of Cleveland. Woolley was near Swanswick where I was then residing, and I held the curacy for eight years, when on my removal to Bath—or rather Bathwick—it was no longer within easy reach, and I gave it up. I had the charge of Langridge parish as well—also within a walk from my house at Swanswick—for nearly two years. This was during the illness of the Rector, who died shortly after the two years had expired; the living then passing into other hands, when my services were no longer required.

Neither of these two parishes had more than about sixty inhabitants, so the work was not very heavy, nor the congregations on Sundays, as may be supposed, very large. The people of Langridge, too, were very much scattered, there being but little appearance of a village at all. On one Sunday in particular in midwinter, at Langridge, I had scarcely any congregation whatever. Snow had fallen all the previous Saturday night, and on starting for church duty it was ankle deep, having to make my own path, as there had been no traffic on the road since the snow commenced. As I approached the church I heard the bell tolling, but saw no one. On entering, there was the clerk only, with the bell-rope in his hand. This was not very encouraging, but I went into the vestry and sat down, in the hope that some others would soon make their appearance. After a while, there came in one man, and having then got my "two *or* three," I marched out in the surplice, took my place in the reading-desk and commenced the service. During the confession, one other man entered, and that was all the congregation I had—two persons besides the clerk. I thought with myself that the best way to meet the circumstances was to take no notice of it at all. I was bound to read the Service, whether the congregation was large or small, and I accordingly did so, omitting none of the prayers, preaching afterwards, and throwing as much energy into my sermon as I was ever wont to do. I did not wish the "two or three" to leave the church with any complaint that they had come there for nothing, or that I had slurred over the service, disregarding the few who *had* come (under the same difficulties as myself), because there were not more.

The above duties availed me but little in a money point of view. The services at Langridge were given gratuitously. The small pay I had for the Woolley curacy I estimated to be at the rate of about nine shillings a week, hardly amounting to the pay of a common day-labourer.

On my removal from Swanswick to Bath, I had the consent of the Rector of Bathwick to visit some of the poor in his parish, which gave me a little further employment in the ministerial way. I also visited Bellott's Hospital, one of the old Bath Municipal Charities, founded in 1511, for the accommodation of poor persons seeking the benefit of the Bath waters. My connection with this Hospital, which I visited weekly (when not from home) for many years, arose from the circumstance of my brother, who lived in Kent, writing to inform me that a police inspector in his neighbourhood, whose health had entirely broken down from repeated attacks of rheumatism, caused by exposure to wet and cold in the discharge of his duties, had got admission to Bellott's Hospital for the sake of the Bath Waters. He asked me to go and visit him. This, of course, I did, and repeated the visit at intervals. After a time some of the other patients in the Hospital asked to be allowed to be present when I came to read to him. Permission being given, this led eventually to my having a weekly short service in the Hospital on a fixed day, when all the patients who were able and inclined would attend, with the matron and any friend with her, the attendance, of course, being quite voluntary. In this way I got together a congregation of about twenty, Churchmen and Dissenters mixed, to whom I read a portion of Scripture, or some religious book suited to the circumstances of the place, ending with a short exposition or address and a few prayers. This practice continued from year to year, until my own health and strength beginning to fail I was obliged to give it up. And with this ended also my ministerial labours, it being between forty-five and fifty years from the time of my ordination.

Medical Profession, with Remarks.

I have spoken above (p. 16) of my early determination to enter into Holy Orders. But though definitely fixed on this, I felt at times a strong leaning towards the medical profession. This is not to be wondered at, seeing that both my grandfather and uncle (on my mother's side) had been such eminent physicians; my mother also knowing much of the subject, and always keeping by her a well-stocked medicine-cupboard, serving for all ordinary cases of illness in the family and household. What turned me from the idea was my own weak health, which would never have stood the fatigue of much practice as a medical man. But I was always fond of the subject. When at Cambridge, I attended the Lectures of Professor Clark, on Human Anatomy, which, with the help of books, gave me a good general knowledge of the human frame and its several organs, carried further by other reading of the best works on Physiology and the treatment of disease. All this proved useful to me in two ways. It enabled me in many cases to deal with sickness among the poor of the parish I had in charge; it further taught me thoroughly to understand the laws of health as applicable to myself.

I was never strong as my two brothers (older than myself) were; but I was born of healthy parents, and had no reason to suppose there was any tendency to disease in any of the chief bodily organs. My case was that of a nervous temperament, which—though distressing at times—wore off, as I believe it often does, with advancing years. All the first part of my life I was subject to sick headaches, which occasionally prevented me from taking the Church duty on Sunday morning; when I was obliged to ask the assistance of a neighbouring clergyman, he giving a single service between his own two services. Twice during the period of my incumbency, my complaint so far increased as to render me unable to officiate for weeks together, though able to attend Church as one of the congregation.

About middle life sick headaches left me entirely, and my health gradually improved. By that time I thoroughly understood my own constitution; and I believe it in great measure due to my paying proper attention to the laws of health that my life has been prolonged so far beyond the proverbial—" three score years and ten." In a few days from the time of my writing this I shall enter my 89th year.

I consider that every educated man ought to have such a knowledge of his own constitution as will enable him to preserve his health,—circumstances allowing and apart from accidents. It is as much a duty to himself to attend to this, as it is a duty we owe to others to do good according as we have opportunity. If a man inherits a morbid constitution, or a tendency to any particular disease, this of course is not his fault, but his misfortune, but even here he should take the more care not to increase that tendency by any acts of imprudence : or he may come within the influence of some epidemic he was not aware of. But these are not the cases now in view. How many ailments arise, not from inheritance or infection, but from preventible causes : from excesses in eating, drinking, and other self-indulgences ; how many from taking cold, young and old alike often exposing themselves to severe winter weather, without any, or quite insufficient, additional clothing, and not unfrequently from a foolish idea of hardening their constitution.

How many, again, shorten their lives by bad habits, as regards study and bodily exercise ; sitting up late at night to read, and then lying abed half the morning, having meals at irregular hours, or no regular meals at all, and taking no regular out-of-door exercise.

I have known some hard workers—in the field or garden as well as in the study,—who would both get up very early in the morning and go late to bed at night, denying themselves sufficient sleep, so that nature was never properly recruited, and the constitution, as might be expected, broke down under so severe a strain. This is, as it were, burning your candle at both ends.

How many, on the other hand, waste their lives in idleness and inaction. I imagine there is no medical man but would allow that every part of the body, to be kept in health, requires exercise, or must do work of some kind. If an organ is disused, it wastes and gradually deteriorates. This applies to the brain especially. Idle men who do nothing,—who from easy circumstances are not obliged to exert themselves for a livelihood, who care nothing for literature, or science, or for any occupation that makes them think ; these are the men who get more and more dull and stupid as they advance in years, and ultimately, in old age, quite childish, if they live to be old at all.

It is equally the same with those who take no *bodily* exercise, who never take walks, or leave the house except perhaps for a carriage drive, which is no exercise at all, and should only be resorted to when a man, from disease or accident, has lost the use of his legs, or when he has to go long distances.

Here, too, there is a wasting from disuse, the calves of the legs fall away, leaving only lean shanks; no proper circulation of the blood is kept up, ending—not unlikely, sooner or later—in paralysis.

In reference to my own health, my habit has always been, as a rule, to attend to two things in particular. One was, circumstances allowing, to exercise, every day, mind and body equally; the other was—after making myself acquainted with the particulars of my own constitution, and as to what suited it best, to lay down rules for ordinary observance and keep to them.

These rules had chief references to eating, drinking, sleeping, out-of-door exercise, and so forth. When habits have been long set up, the body, as it were, grows to them and regulates all its functions accordingly. Of course they should be good habits, as generally allowed, in the first instance; but the bodily organs are so jealous of any interference, when they have been long accustomed to certain ways of living, that even bad habits, which as a rule are to be avoided, should not be too hastily exchanged for others when they have been practiced for any length of time, and some perhaps, in old age especially, when the body must to a certain extent be humoured, should not be interfered with at all.

In a Thunderstorm.

On May 29, 1859, when resident at Swanswick, I was out for the day, and exposed to the full force of the heaviest thunderstorm within my recollection, except two, both these being characterised by the fall of very large hailstones. Fortunately there was no hail in this case, or the consequences might have been more serious—if not fatal; but there was a continuance of rain in torrents, with incessant thunder and lightning, for many hours together, the whole of which I had to face without umbrella or shelter of any kind. The circumstances were as follows :—I had arranged with my late friend Broome, who lived at Batheaston, to walk to Rudloe, about 6 miles from Bath, and much nearer to him than to myself;—he had also formerly lived at Rudloe himself, and knew a shorter way to it, across the fields, than by the road. Our object was to get to a particular wood, where he used to find a rare plant, for which no other locality was known in the Bath district. On our arriving at the gate leading into the wood, Broome said he had been once warned off by the proprietor, and he did not like to venture in himself; but he directed me to take a cart-way that went a certain distance into the wood, and after a while I should probably find the plant growing in the underwood by the roadside. Having got the plant, I was to go straight on, the road ending in a path which led to a mill where he would wait for me.

While on our way to Rudloe, in the forenoon, I had

noticed the appearance of the sky as rather threatening, and not unlike that of an approaching storm, and sure enough it was so. I had only just reached the spot, where as I supposed the plant might be found, when a loud clap of thunder warned me of what was at hand, heavy rain at the same time beginning to fall. To be in a wood in a thunderstorm where every tree was a conductor, was not I considered a very safe retreat. So after making a very hasty search for the plant which I could not find, and not having much time or inclination to look about for it under the circumstances, I deemed it prudent to get out of the wood as soon as possible. Accordingly, I followed up the road that was to lead to the path, which Broome directed me to take in order to get to the mill where he would be waiting for me. To my dismay, when I got to the end, I found the underwood closing all round, and no path at all. It was probably many years since Broome had been there himself, and all was now altered. I then looked to see if there was any other outlet from the wood, but could find none. There was one indeed which I tried, but it brought me on to the wide down, without a track to guide me as to my way home. I had no idea, indeed, where I was. Nothing, evidently, remained to be done but to retrace my steps to the gate by which I had entered, and then take the most likely road to Box. After getting there, I knew my way home at once. I was quite wet through before leaving the wood, so I thought no more about the rain which kept falling in torrents, but the thunder and lightning was fearful and incessant.

I had gone a certain distance on the road when I saw a labourer, who had been ploughing, hastening with his horses out of the field by a gate just where I was standing. I hurriedly asked him whether I was on the right road for Box, but the words were hardly out of my mouth, when a crash of thunder and lightning, quite close to us, so frightened the horses that off they galloped with the man after them, before he had time to reply to my question. So I was again left to myself amid the raging elements. I could only walk on in hope that the road I had taken would bring me to Box, which it did at last, but not till after trudging a long distance, much exhausted, and under apprehension of being struck any moment by the lightning which was playing around me all the way. On getting to the village I rushed into the first public-house I came to, and begged to have a glass of hot brandy and water as soon as possible. I dare not stop to rest, or even sit down for a minute—I was so very wet—lest I should get a chill. But after the brandy, which much revived me, I felt I had energy enough to complete my journey home. I had still three miles or more to go, the storm continuing for the first part of the way, but gradually going off as I got nearer to my own house, which I did not reach till near seven in the evening. The storm had commenced about noon, and thus had lasted nearly six hours.

On getting home, I took off my wet clothes, had some tea, and went to bed immediately. The next day was Sunday, and I felt quite well enough to go to Church, and even to address some Sunday School children. Nor did I feel any ill effects from my wetting and fatigue on Monday; but the third day after (Tuesday) I had uneasy sensations in my head, and a feeling of depression which obliged me to keep to the house, sitting perfectly quiet and doing nothing. I sent to Bath for my medical man, who said the wetting and exhaustion over so many hours was a serious thing, and had brought on a low fever. Tonics and complete rest were the only remedy, so under his advice I kept entirely to my arm chair. I had no pain of any kind. It seemed rather a positive enjoyment to be allowed to sit perfectly quiet, seldom even opening a book, and doze when so inclined. This continued for many weeks, the doctor visiting me from time to time, and it was not till near Autumn that I fully recovered my strength, and was able to resume my ordinary occupations.

The distress and difficulties I met with that fearful day, to say nothing of the danger I was in from lightning, left an impression that can never be effaced from recollection. I have always considered that the brandy and water I got at Box was in great measure the saving of my life; but for it, I really think I must have succumbed before reaching home.

Isle of Wight.

It was in October, 1849, that my wife, maid, and self, left the Vicarage House, at Swaffham Bulbeck (my wife never to return to it), and started on our journey to the Isle of Wight, as alluded to above. Ventnor was chosen as, on the whole, the most desirable locality in which to take up our quarters. We accordingly hired a house there, and remained in the island eight months. The following winter proved to be a very severe one, more so—according to the native inhabitants—than any that had occurred for many years back. The snow lay deep on the ground for several weeks in succession. Cold, however, as it was, we felt sure that we were better off than had we remained in Cambridgeshire, where the weather would have been certainly more severe still.

As the spring gradually dawned upon us, I became impatient to get more abroad and commence my rambles in the neighbourhood; as well to see the scenery (so different from that of the Cambridgeshire fens), as to explore its Natural History. I saw a good deal of the island during our long stay, but scarce any part of it is more attractive than the "Undercliff," stretching along the south coast, from Dunnose to Blackgang, or further; Ventnor being nearly at the eastern extremity. The chief romantic scenery was consequently close at hand, as well as some of the best ground for botanising. On the cliff, near Steephill, I gathered specimens of the rare wild stock *(Matthiola incana)* in full flower, but it was very sparingly to be found, and is not unlikely extinct there at the present time. In corn-fields, above the Undercliff, the pretty pheasant's eye *(Adonis autumnalis)* and the purple cow wheat (*Melampyrum arvense*) were in abundance, plants met with in few other parts of England. On the whole, I made considerable additions to my Herbarium (now in the Jenyns' Library), in which they may be all seen.

Bonchurch, at the eastern end of the Undercliff, and not far from Ventnor, is a very pretty spot. The old Church, with its churchyard, are objects of much interest; the latter, especially from the circumstance of the Rev. William Adams, Author of "The Shadow of the Cross" and other allegories, having been buried there. "To commemorate this, a raised cross has been placed over the stone which covers his remains, in such a manner that its shadow always falls on his grave, and he may be truly said to rest under the Shadow of the Cross." The Church is in rather dangerous proximity to the sea, which is slowly encroaching upon the cliff, and threatens to swallow up the Church eventually.

The new Church, a very elegant structure, and well arranged for congregational purposes, had only been completed and consecrated the year previous to that of my coming into the neighbourhood. In this Church I often officiated during my stay at Ventnor, assisting the curate, the Rev. Fielden, either in the desk or the pulpit. Archdeacon Hill, of Shanklin, I think, was the Rector, but confined to his house at that time by a broken leg. On one occasion I was asked by Mr.

Fielden to preach at Bonchurch, in order to keep another party from the pulpit. It was at the time of the celebrated Gorham controversy respecting Baptismal regeneration. Dr. William Sewell, of Oxford, well known for his extreme tractarian views, had a sister living at Bonchurch, whom he occasionally visited, and at whose house he was staying at the time alluded to. He was in the habit—whenever he came to Bonchurch—of asking Mr. Fielden to lend him his pulpit, which ordinarily the latter was quite willing to do. But this was not an ordinary occasion, and Mr. Fielden, who was a man of moderate views, and who attached himself to no particular Church party, feared that if he let him preach just then, he would introduce controversial matter into his sermon, and stir up unpleasant feelings in his congregation. So he came to me beforehand and asked if I would preach for him the following Sunday, in order that if applied to by Dr. Sewell he might say his pulpit was engaged. I could not but agree to do so, though it was rather a nervous affair to me, who disliked all controversy, especially as the adversary in question would be one of the congregation. I was obliged to be extremely cautious as to what I should say in my sermon. It was impossible to avoid the subject altogether; it was the talk of everybody—but I said very little about it, and nothing that I thought could be laid hold of or give offence to either party.

We sometimes went to the little church of St. Lawrence, standing on the summit of a hill, and not very far from Ventnor. It has the reputation of being the smallest church in England, and so I think it must be. I one day roughly measured with my walking-stick the breadth from north to south wall, and found it scarcely more than five lengths of the stick, or about fifteen feet. The chancel had been lengthened a few years previous, and as the village population was, I think, said to be increasing, it is not unlikely the whole building may have been enlarged since I was there.

The service at the time, and the arrangements connected with it, were very peculiar. There was a wooden structure with a desk, enclosed all round below as in a pulpit: this served for prayers and preaching alike. At the end of the prayers the clergyman gave out the hymn and commenced singing it, at the same time gradually lowering himself until he was out of sight—still singing—when he changed his surplice for the gown; the operation over, he reappeared at top and finished the hymn. Such an adjustment of things was enough to provoke a smile; and in truth, a little girl, six or eight years old, with her mother, in a pew close adjoining the one I sat in, was so amused as to call out to her mother—loud enough to be heard by others near—" Ma! ma! go down white man; come up black man!" All this is probably changed by this time.

We made several pleasant acquaintances in Ventnor and the neighbourhood. First and foremost was Dr. George Anne Martin, the physician of the place, and author of an excellent work on "The Undercliff," in which its history, climate, and natural productions, are all dealt with in such way as to make it a valuable guide-book for strangers. It was published the same year I came into the island, and proved of great use to me.

Dr. Martin's remarks on the climate of the Undercliff are based upon Meteorological Observations, which he had carried on himself at Ventnor for many years. Having been myself an observer in that way a great part of my life, we had naturally much to say to one another on the subject. It so happened that the Meteorological Society was first proposed and started that same year in which we were first brought together. Both he and myself received a circular inviting us to join. He wrote to me on the occasion, asking my opinion and what I thought about it. For myself I declined; there being no names down known to me as meteorologists, or indeed as men of science in any way, and I thought it very doubtful if the Society came to anything, or would ever do work of value or importance. On having my answer, Dr. Martin—who was much of the same opinion as myself—declined also. Our forecasts, however, in this matter, proved to be quite in fault. The new Society gradually took root and flourished; its publications have been valuable and numerous; it has even obtained the patronage of Royalty, bearing now the title of "Royal Meteorological Society."

It was at Dr. Martin's house, at an evening party, that I first met and got acquainted with Dr. Bromfield, who was then working out the Botany of the Isle of Wight, and getting materials for his "Flora Vectensis." This valuable and laborious work he never lived to publish, or thoroughly to complete. Dr. Bromfield is said to have had a passionate love for travel; and it was after an excursion to Egypt in 1850, while prolonging his tour into Palestine and Syria, that he was cut off by a fever at Damascus. The materials he had got together for the above work were placed in the hands of the late Sir William Jackson Hooker and Dr. Thomas Bell Salter, who edited the book under the aforesaid title—no mere catalogue of species, but a thick octavo volume, entering into all the circumstances connected with the local history, habits, and characters of each species. The book was published in 1856. A copy of it is in the Jenyns Library, kindly sent to me by the author's sister.

Besides Dr. Martin; we saw a good deal of Mr. Fielden, of whom I have already spoken, and his pleasant wife and daughters. A few other persons at Bonchurch and Ventnor we met occasionally. Altogether we much enjoyed ourselves in the Island, and— but for the illness of my wife, which brought us into it— I should look back upon it as one of the most agreeable "Chapters" in my life.

We quitted the Island in July, 1850, and after a visit to my wife's friends in Gloucestershire came to Bath in the autumn of that same year.

Meteorology.

In the 8vo edition of White's Selborne, published by Markwick in 1813 in two volumes, there are, at the end of the second volume, some "Miscellaneous Observations on Animals and Weather, with a Calendar of Periodic Phenomena in Natural History," extracted from White's Diaries, none of which appeared in the original 4to edition published by White himself.

On reading these over, whilst yet at Eton, I conceived the idea that Observations of a similar kind, carried on for some years, might furnish matter for two separate works, one containing Observations relating to Natural History; the other, those relating to Weather Phenomena and Meteorology. This led in after years to the publication of my two works—"Observations in Natural History, etc.," published in 1846, and "Observations in Meteorology," published in 1858. It is in reference to the latter subject of *Meteorology* that I would now speak.

I added this subject to my other scientific studies very early in life. In fact, it was at Eton, after reading White's "Observations," spoken of above, that I felt a desire to take an account of the weather, and commenced observations of my own—of course, rough and after school-boy fashion. I well remember running up Eton Town Street, every day that I was able, to look at a Barometer hanging up in Rogerson's shop, the cutler so well known to all Eton boys of that day, near Eton Bridge.

After leaving Eton, the observations were continued—first at Cambridge, then at Bottisham Hall, afterwards at my Swaffham Bulbeck Vicarage,—but were of no scientific value, from the irregularity with which they were carried on, and from the imperfect instruments I employed.

In 1830 I took the matter up more seriously. Well settled in my new Vicarage House, I procured from Newman in London, the great Philosophical Instrument Maker of those days, the best Barometer that could be constructed, besides Thermometers and Hygrometer, and made observations regularly twice a day, at 10 a.m. and 10 p.m., only interrupted when away from home. This continued until the time of my leaving Swaffham Bulbeck in 1849, a period of 19 years. After settling down again at Swanswick, near Bath, the results of the Cambridgeshire observations were gradually worked out and embodied in a volume comprising the chief features of the Climate of that county, under the title and in the year already stated.

One part of the subject which greatly interested me was that relating to the Fogs and creeping Mists which so particularly characterise all the low grounds—near to or on a level with the Fens.

A fairly large grass meadow in the front of the Vicarage at Swaffham Bulbeck, was the field of my chief observations on these phenomena, and when circumstances seemed likely to favour the formation of such mists, I used to place—ready beforehand—at the further end of the field a chair and small table supplied with thermometers, hygrometer, and note-book. I thus took account of all that occurred from their first appearance until their disappearance, if they did not continue all night; the results I need not state here, as they are given in my book.*

In these researches I was much assisted by my village schoolmaster, a very intelligent young man, and of rather superior abilities to most others in his situation. Indeed, he learnt to take such an interest in Meteorological work that, after a time, I instructed him how to make the daily observations of Barometer, Thermometer, etc., at the Vicarage House, and to keep up the Register, when I went from home.†

In 1864, many years after leaving Cambridgeshire, and whilst living at Darlington Place in Bath, where I still continued my Meteorological Observations, the British Association came to Bath to hold their Annual Meeting—never having visited it before. Being anxious to seize the occasion for saying something about the Bath climate, and having little of my own to bring forward on the subject, I applied to one or two other observers, who had lived in Bath much longer than myself, and who had kept registers—one of them especially for a considerable term of years—to favour me with the results of the same, as materials for a paper, which after putting together I read to the physical section, and which seemed to interest the hearers. I discovered, however, afterwards, that the observations upon which this paper was based were not altogether

* See "Observations in Meteorology," pp. 197—210.

† This young man, in after years, rose to being a Schoolmaster at Bury-St.-Edmunds, of high repute, and much patronised by the Clergy.

trustworthy; and bringing the subject under the notice of the Bath Literary and Scientific Institution, I prevailed with that body to consent to having a little Meteorological Observatory set up in the Institution Gardens, all the arrangements connected with the erection and the instruments to be placed in it being left to myself. This was with a view to beginning—and keeping up for the future—a Meteorological Register, under the conditions necessary for insuring its utility and permanent scientific value.

The observations were commenced in March, 1865, being made by the Librarian of the Institution, who entered warmly into the matter; and at the expiration of ten years, *i.e.*, in 1875, the results were worked out by myself, forming the subject of a long paper read to the Bath Natural History Field Club, in which I embodied all the particulars I could get from others, in addition to those derived from my own experience,—relating to the Bath Climate. I caused to be printed, for my own private distribution, 100 separate copies, which I distributed amongst the best known Meteorologists in this country, and other men of science, as well as Scientific Institutions; giving copies also to friends and Medical men in Bath.

Another decennial period was completed in 1885. I have lately brought together the results of the whole twenty years, which—it is to be hoped—there being no break in the observations—will convey a fairly correct impression of the climate of Bath, while at the same time it makes a finish of all I have to say on the subject of Meteorology.

Ray Commemoration.

I was present at the great dinner at Freemason's Tavern, on Nov. 29th, 1828, held in commemoration of the 200th Anniversary of the birth of John Ray, the celebrated Naturalist. There was a very large gathering of Naturalists and men of science generally on the occasion, Davies Gilbert, the then President of the Royal Society, being in the chair. A long list of toasts had been drawn up, each of which in its turn was prefaced by a few suitable words from Davies Gilbert, who was quite up to the mark; he being not only well conversant with the chief branches of science, but himself a ready and fluent speaker.

I had a seat assigned to me between Kirby, the well-known Entomologist, and Henslow. Both had to speak. Kirby responded to the toast of "The Naturalists of Great Britain and Ireland," and Henslow to that of "The University of Cambridge."

Kirby, much advanced in years, was very nervous when his turn came, and while speaking kept his right hand on the upper bar of my chair to steady himself, causing it to shake so much that I was in a tremor the whole time of his speaking. However, he acquitted himself very fairly. Henslow's speech came on later. He said, in reply to an allusion from the Chair as to Ray having been expelled from the University on account of his religious opinions, that the University had done all in its power to wipe away that great act of injustice, his bust being ranged with those of Newton, Boyle, Barrow, Willughby, and others, and that his spirit still lived there. Ray was the greatest Naturalist of his age.

Buckland had to speak on the side of the Geologists. His speech was full of eloquence and abounded in wit and humour, rather bordering, however, in one part of it on coarseness and buffoonery. I remember but little of it except his saying—when speaking in praise of Hugh Miller, author of "Testimony of the Rocks" and other works,—that "he would give his right hand to write like that man." Buckland was more of a ready speaker, who had always plenty to say, than "a ready writer."—It is said that he was much assisted by his wife, who was a very intellectual Lady, in his Bridgewater Treatise.

ANECDOTE OF DR. BUCKLAND.

One summer's day, I think in 1844, I was surprised by a call from Dr. Buckland at my Vicarage house, at Swaffham Bulbeck. He had come from Cambridge and much wanted to see the Reche Chalkpits, about two miles from my house; well-known in that neighbourhood from those pits supplying the greater part of the Clunch so generally used about there for building purposes.

Clunch is the local name for the *Lower Chalk* of Geologists, which is well developed in those pits. After Buckland had satisfied his geological curiosity, we strolled into the fen close adjoining, and on finding some of the poor digging *turf*, as it is called there, the chief fuel used for fires,—elsewhere more generally termed *peat*—he took a great interest in the matter, asked me to collect all the facts and circumstances connected with it, and to put them into the form of a paper for the British Association. This I accordingly did, reading a paper on the subject to the Natural History Section at the Cambridge Meeting of the British Association in 1845.*

On the return walk to my Vicarage, Buckland stopped short at a cottage in the village, where he saw some turf piled up outside the door dry and ready for burning. Taking a piece in his hand, he stepped in and asked an old woman inside the house what was the price? Bewildered at the sight of the would-be purchaser and the article he wanted, she scarcely knew what to say, whereupon Buckland gave her fourpence, at which she was greatly pleased (its real value being about a farthing), and, putting the turf in his pocket to add to his omnivorous collection of things having the most remote bearing upon Geology—walked off,—leaving the old lady staring and puzzled to think who the gentleman was and what he was going to do with it.

* See the paper entitled " On the Turf of the Cambridgeshire Fens," in the " Report " of the Meeting for that year, *Communications to the Sections, p. 75.*

THE BRITISH ASSOCIATION.

The British Association was started by a few kindred spirits in the line of scientific research, in 1831. It met that year for the first time at York, where it met again in 1881, fifty years afterwards, to celebrate its jubilee. About 350 assembled at the first meeting, most, however, if not all, of those whose names are known to science have since passed away. I was not at that meeting myself, but I joined the Association very shortly afterwards, and made my first appearance at the second meeting held the following year at Oxford. I remember well being one of four, who joined in hiring a carriage to post from Cambridge to Oxford on the occasion, the other three being Henslow, Clark, Professor of Anatomy, and Bowstead of Corpus College, afterwards Bishop of Lichfield.

The President of the British Association for that year was Buckland, and the President of the Natural History Section, to which of course I chiefly attached myself, was Mr. Philip Duncan of New College—not better known in Oxford than in Bath, where he resided all the latter part of his life, and became one of my most esteemed acquaintances till the time of his death at the advanced age of 93.

I was glad to meet again at this Oxford gathering of notabilities—not in science only, but in rank and other ways—my old schoolfellow, the Hon. George Howard, as he was then, being afterwards first Viscount Morpeth, and next Earl of Carlyle, of whom I have already spoken.

For many years subsequent to the Oxford gathering of the British Association, I attended the Meetings very regularly, often arranging my summer tours so as to fall in with the different places at which it met. In this way I got to know a large number of the Naturalists of that day, especially those locally connected with the several Towns at which it assembled;—and we associated much together and enjoyed the week's work. Gradually, however, death laid its hands upon many of those I knew best,—my own attendance became less regular,—and at this present time, 1887, scarcely one survives of those with whom I was so familiar, and with whom I companied as following up the same pursuits as myself. Other men have come forward in the room of those departed; and they are doing good work in their respective lines of research. But they are mostly strangers to myself, and I feel that my own work is done. There is a time for all things;—and a time especially to withdraw from those studies and pursuits which require much thought, and energy of mind and body.

THE BATH FIELD CLUB.

"The Bath Natural History and Antiquarian Field Club"—now numbering 100 members or more, though only two remain of the *original* members–was founded by myself on February 18th, 1855. Its object was to make excursions round Bath, with the view of investigating the Natural History, Geology, and Antiquities of the neighbourhood.

Having given a detailed account of the circumstances under which this Club was first started, and some account of its doings, in an Address I read to the Club on the occasion of the sudden death of my old and much-esteemed friend, Mr. Broome, who—from our close companionship in all that related to its first establishment—might almost be called a joint founder with myself, I need not repeat them here.

The Club has gradually increased its numbers and extended its work. The best signs of its progress, perhaps, are to be found in its "Proceedings," already alluded to. I am no longer able to join in its walks or to take part in its evening meetings, but, in the hands of others, I trust it may still go on and prosper, pursuing its even course without check, or the falling off of those on whom it leans for support.

The World Slowly Improving.

No one who knows anything about the growth and progress of Christianity and the spread of civilisation can doubt for a moment that the world improves from century to century. Yet, there are not a few narrow-minded persons who—measuring all things by their own rule—will not allow there can be any general advance towards the goal which the Gospel sets before us, unless the "race"—to use Paul's metaphor (I. Cor. ix. 24)—be run upon the exact lines they lay down for themselves, and in accordance with their own views as to what is right and what is wrong.

There are even some who don't care that the world *should* advance. They would have things remain as they are, and they grumble at all innovations—new customs, new ways of doing things. Straining the text in St. John—"The whole world lieth in wickedness"—they think it is always to do so; that it is of no use trying to mend it, and the sooner they themselves are out of it the better.

Persons of this morbid disposition would do well to listen to what Bacon says on the subject. Bacon likens innovations to medicines. They are introduced to correct what is old or worn out in things generally considered. He says:—" Every medicine is an innovation, and he that will not apply new remedies must expect new evils; for time is the greatest innovator; and if time alters things to the worse, and wisdom and counsel shall not alter them to the better, what shall be the end? Time moveth so round, that a froward retention of custom is as turbulent a thing as an innovation; and they that reverence too much old times are but a scorn to the new."*

It is of little use arguing with the persons above alluded to. Complaints about the wickedness of the world do no good. Its reformation, as a whole, can only be carried out on such lines as Providence has laid down. None of us, singly, can do much for its improvement. Yet, if all would simply give themselves to the full discharge of what is in each man's power, according to the state of life in which God has placed him, the result would be very considerable. The world would then be manifestly better, and its wickedness lessened. Eventually, great ends may come about by the accumulation of small attempts to do what is right and good. If we do but give a cup of water to drink to one of Christ's disciples, we are told we shall not lose our reward. I remember a remark of Dr. Whewell's, a few years after he had taken his Bachelor's degree at Cambridge:—" Herschell and I," he said, " have determined to leave the world a little better than we found it." This was in allusion to an attempt, on their part (in which so far as I recollect they succeeded), to get the University to adopt certain mathematical works, by distinguished men, on the Continent, of a higher grade than those in general use by the College students, these last being behind the age. Such a

* Essays, xxiv. of Innovations.

change, in the first instance, effected the University alone. But who can say what it may not have led to? Who can set limits to the good results that might accrue from any improved teaching in a University, where young men from all parts of the kingdom are brought together for educational purposes? If the above determination were made by all, how soon would the world be reformed; but alas, it is left to the "few" to make; the "many" care nothing about it.

But to come to the experiences of my own life in connexion with this subject. I have often said to myself, in reference to our Lord's comparison (Matt. xiii., 33), how slow the "leaven" works. But that it does work, that the world does improve, however slowly, under the influence of our Lord's kingdom, is clearly proved by what a close observer can recollect of the past to set against what he sees and hears at the present day.

Born in the last year of the last century, and yet living now—when the century succeeding that one has nearly approached its last decade—my life has run, as it were, parallel with the age, and the circumstance of its doing so enables me to contrast the state of things in early days with things as they are at the present time.

I say nothing about the advance of the arts and sciences, which would be quite foreign to my purpose. I would speak of things more personal. I would allude, first, to the altered and greatly improved tone of conversation in society—say, at a dinner party. Those who are only acquainted with the men of the present generation, or not much before it, would be shocked to hear what I had sometimes to sit and listen to at the tables of country gentlemen. I may even say at my own father's house, after the ladies had withdrawn, matters of the grossest indecency would be discussed, especially if anything had occurred, or been mentioned in the public papers, to lead to the subject, with such freedom and particularity as might be necessary in a court of law, but quite unbecoming the conversation of a party of gentlemen.

Once, at a Curate's table, in a parish not far from my own, the Curate himself brought on a conversation (suggested by his own unsanctified heart) about such things, as the Apostle says, "it is a shame even to speak of;" not very creditable to his profession—indeed, so disgusting, I thought I must have left the room. Who, among the upper classes, would ever talk of such things at the present day—above all, in a company consisting mostly of clergy?

So, also, with regard to religion. There were gentlemen—and gentlemen of rank, too—in those days, who, though Christians by name, and perhaps Church-goers from habit, in social conversation, would sometimes mock at its forms and rites in the most irreverent way, ending their remarks with a coarse laugh.

I imagine this seldom happens now. There may be many men semi-infidels, who write against religion, or preach down some of its most characteristic doctrines, but few who, in genteel society, would turn it into ridicule. They would either keep their thoughts to themselves, or enter upon a fair argument with their opponents.

I would now speak of the clergy generally, and their profession. I have stated above how many parishes there were in the neighbourhood of Cambridge, when I first took orders, without resident clergymen to look after the people. I should think there is now hardly one such. Everywhere throughout the country, residence is the rule, non-residence the exception. Pluralities also are much less frequent. They are not altogether abolished, nor is it perhaps desirable they should be; for, in some cases, if two livings are near together, they may greatly help each other. I remember a case in Suffolk formerly, where there were two parishes a mile or two apart, one of them richly endowed, with a small population, held by a friend of mine; the other, of very small value, with a large population, but no resident clergyman, nor indeed any house for him if there were one: this last was a college living, and ought to have been more considered by the patrons. The Rector of the first parish, above mentioned, said to me one day,—" If they would give me that living, my better-endowed one would enable me to place a resident curate there, and, between us both, we could work the two parishes together well." Perhaps there are not many such cases; and far more frequent were the cases at that time, in which the money-getting clergy would lay their hands on as many pieces of preferment as, by any stretch of the law, they could lay hold of. One clergyman, I remember well, was not only a Canon of Ely (Prebendary in those days), but held, together with his Cathedral stall, so many livings in Suffolk, that he was often called in joke the *Rector of Suffolk*.

And many were the cases in which clergymen, without being pluralists, broke away from residence on the one living they had, on the most insufficient grounds. There are not a few, it is to be feared, even at the present day, who almost, if not quite, break away from their profession altogether. Some, from weariness of work and constant repetition of the same round of duties; others—where there are but few of their own standing in the neighbourhood to consort with—from want of society. The first evidently never had their heart in their work, and chose a wrong profession in the first instance; or they took to it from wrong motives, and other considerations than those of wishing to do good in their generation and to "leave the world a little better than they found it." Those of the second class would do well to read Paley's sound advice to young clergymen in one of his ordination sermons. He says, " Learn to live alone. It is impatience of solitude which carries you continually from your parishes, your home, and your duty, makes you foremost in every party of pleasure and places of diversion, and puts you out of humour with your profession."

And here we see the advantage of having some secondary occupation to turn to at leisure hours, and so shutting out all feelings of ennui—a fondness for books and reading, or a taste for some branch of science, such as I had for Natural History. Time never hung upon my hands; I was more often racing against it and wishing the day longer.

One severe winter, when the country around was covered with deep snow and the family were away from Bottisham Hall, I remember being five weeks without conversing with any one except my servants and some of my parishioners. But my pursuits never allowed me to be otherwise than happy and contented.

I consider it, moreover, due to my love of work and constant occupation of the mind, alternating with out-of-door exercise when practicable, that I have lived to my present great age, without any serious interruption of good health or stop to my daily pursuits.

Leaving now the consideration of the lower clergy, I would say a few words in reference to the higher order of Bishops. Here, too, there has been a manifest improvement. The Bishops in the early part of this century—in the days of my first ordination—were as different from the Bishops of these days as can well be imagined. We now see the Bishops moving actively about their dioceses, attending to the affairs of the different parishes in them as required, listening to their several wants, ready to help them with their advice, or in other ways, according to circumstances; to say nothing of diocesan conferences, etc., in which both clergy and laity are invited to join for the benefit of each other's judgment and opinions.

Formerly, some of the Bishops were heads of houses in the Universities, where they principally resided; it appearing as if they had exchanged the care of a diocese for that of a college. There were two such at Cambridge when I was an undergraduate. In my younger days they all wore wigs, and they were called big-wigs.

Kaye, Bishop of Lincoln, by whom I was ordained priest, as stated above, was the first Bishop who did not wear a wig. He was only 37 when first raised to the Episcopate, and it was thought the wig might be dispensed with in the case of so young a Bishop. But though young, he was a very learned man, a senior wrangler, as well as first Chancellor's Medallist, and Regius Professor of Divinity at the time of his consecration. From his time wigs became less and less frequent, until they disappeared altogether.

In regard to the intercourse between the Bishops and the other clergy (I speak here only of the diocese in which my own parish was situated), it was almost entirely restricted to occasions of necessary business, when the Bishop carried himself in a stiff, formal manner, more like that of a schoolmaster than a chief pastor of the Church; his business letters, too, assuming very much the character of a lawyer's letter to an opponent.

In truth, the Bishop rarely had any intercourse with his clergy (his own chaplains, of course, excepted) but at a visitation, which, strange to say, in the earlier years of my ministerial work was mixed up with a confirmation, the two things being held together and only at wide intervals of time to suit his lordship's convenience. The arrangements, in consequence, those especially connected with the Confirmation, were not very decorously conducted. A large number of parishes being brought together, and the children, of course, very numerous, the usual prayer—ordinarily repeated in the case of each child, or that of two children at most—was said but once, and made to serve for a whole railful of candidates, the Bishop's hand passing all the while rapidly over the heads of the children, so as that the prayer should just come to a finish with the last child. On the occasion of one Confirmation, when several children from my own parish were present, the church was the scene of the most disgraceful confusion and disorder. A large number of children had been brought to the place in open waggons at a slow pace and from such distances that they had to start at six in the morning or earlier, in order to arrive at the church in time. The children, when they got there, were tired and weary, some half asleep, others hungry and wanting food, and, when all packed together in a large pew, like sheep in a pen, things occurred such as may well be imagined, but quite unmentionable here. Throughout the church, the crowd and disorder was so great—it will scarcely be believed—that I was never able to get near my own flock during the whole service, and at the close of it was almost in despair, not knowing whether they had been confirmed or not, when fortunately I fell in with my churchwarden, who assured me it was all right.

They say, "when things get to the worst they mend," and so in the present instance. The arrangements for the Confirmation just spoken of were so disgraceful, so unseemly and irreverent in the extreme as connected with a church service, that a vigorous pamphlet on the subject was written by one of the clergy and widely circulated in the neighbourhood to call public attention to it. The publication took effect. A different Bishop officiated at the Confirmation next after the one above, and from that time things generally improved. The Confirmations were held in more churches, so that the children had not so far to go and were not so crowded in the church itself, until in the year before I left Swaffham Bulbeck there was a Confirmation in Bottisham church, the parish next to my own, where I took my sixty candidates, and "all things were done decently and in order."

To an Archdeacon's visitation I never was summoned during the whole period of my residence in Cambridgeshire, nor did I ever see an Archdeacon but once, when he unexpectedly appeared in my church, having heard that I was doing something to it, to see what I was about. I told him I was repairing the rafters, which in some places were rotten from age, and scraping off the coat of whitewash, which, laid on year after year, had quite concealed some of the more decorated features of the building. He seemed satisfied, and said but little, walking off, after my remark that I never had the honour of receiving an Archdeacon in my church before. As for an Archdeacon's charge, I never heard one in my life till after my removal to Bath, and my having a curacy in that neighbourhood. Rural Deans, too, were quite unknown in those days in Cambridgeshire—at least, I never heard of any.

And now, at the end of sixty-five years, dating from the time of my own ordination—perhaps I may be listened to if I say a few words respecting the Church as a profession, or, indeed, as regards the choice of any profession for a young person soon to enter upon the world. Do not begin by saying, you wish him to be this or that, but endeavour in the first instance to ascertain as far as possible what he is best fitted for, and what he himself seems most disposed to consider and think about. A leaning towards some particular occupation will often show itself very early in life, and it may be distinctly brought out by frequent conversation on the subject or matters relating to it; also, by varying the conversation as circumstances arise, and—what is perhaps most to the purpose—by taking a young person out into the world, where he may see for himself what is going on in connection with the work and business of the several professions and occupations man takes to, he will have his mind opened and enlightened more than in any other way. Boys of a certain age are mostly very inquisitive as to the things they see and hear about, and it is not difficult in many instances to judge at once what their inclinations lean to, and what most engages their mind or attracts their attention.

But on no account, if the ministry of the Church be thought of, press or encourage him to enter upon it because of a family living that may come to him, or from any other pecuniary considerations. He must enter the ministry from pure love of the work it entails. He must take pleasure in its religious services, in visiting the sick and poor, teaching in the parish school, etc., according to the position he is placed in.

As a curate, let him be content to remain as such till higher preferment is offered to him, which, if he conscientiously discharge his appointed duties, he may look for in course of time. And if he comes to be a beneficed clergyman, never let him be chargeable with non-residence beyond what the law allows him, or if compelled to be non-resident from ill-health or other unavoidable circumstances, let him remember his duty is at once to resign his living.

Surely, all must allow that the ministerial profession is not one to be taken up lightly, nor till after long reflection on its responsibilities. It engages a man to dedicate his life to the service of the Church and the care of the souls committed to his charge. Yet there are some clergymen who seem never to have learnt this lesson, who keep to a mere routine of enforced duties, or they escape from ministerial work as much and as often as they can. They allege, perhaps, as an excuse for non-residence, imaginary ill-health, and the necessity for going abroad, or to other places in our own country, shirking work, but still taking the emolument of the living, as if "the labourer" were "worthy of his hire," though, in fact, he does no labour at all.

But, passing over these exceptional cases, it is beyond dispute that the clergy as a rule are far more attentive to their duties now than formerly, far more generally resident in their respective parishes; while the immense increase of schools, clubs, and other institutions for the benefit and improvement of the lower classes, compared with what was the state of things in the early part of this century, is most remarkable. And taking this in connection with the multiplied establishments all over the country—one might say, throughout the civilized world—for one good purpose or another, they together sufficiently testify to the world as moving on, however slowly, towards the moral and religious improvement of the human race.

It is mainly left to ourselves, under the directing influence of the Holy Spirit, to reform the world. And the reformation advances, " here a little and there a little," in spite of all delays and difficulties. The leaven works and will continue to work "till the whole be leavened." We must not be impatient. No reforms of any lastin gutility are otherwise than very gradual. In the essay before referred to, Bacon says, "Men in their innovations should follow the example of time itself, which indeed innovateth greatly, but quietly, and by degrees scarce to be perceived." We read also "that one day is with the Lord as a thousand years, and a thousand years as one day." For myself, I have faith to believe that all wrong shall be righted in the end. And as with the wrong so with the wrong-doer— "Every plant which My heavenly Father hath not planted, shall be rooted up."

PARTY SPIRIT.

My father was an old-fashioned Tory (the word Conservative in politics was then unknown) and a strong party man, as most of the country gentlemen were in those days in that part of England in which he lived. For myself I have made it a rule through life to be of no party, and to belong to no party, in either Church, religion, or politics. I think for myself; and, keeping as far as I can, *au courant* with the literature of the day, at least as regards Church matters and subjects allied to it, or subjects bearing upon it, as science, which is pressing hard upon religion at the present time, I form my own views of things, which may or may not be the same as others.

Sarah Coleridge has remarked that "just so far as we become absorbed in a party, just so far are we in danger of parting with honesty and good sense." *

Thousands of persons attach themselves to a party at the instigation of friends, or because their fathers belonged to it, without enquiry, not caring to know anything as to the principles upon which the party is formed, or where the truth lies, or being able to give reasons for belonging to that party rather than to any other. They may, indeed, not have the faculties of thought and judgment requisite for determining the question. Their rank in life and insufficient education may forbid the attempt. But, then, they should stand

* *Memoir and Letters of S. Coleridge.*

aloof from all party demonstration. They may think as they please, and if obliged to side with one party or another, as, say, on the occasion of a general election, they may—indeed, must—necessarily act according to their thinking, but they should proceed no further. At best they take their principles on trust; they know but little of those from whom they have received them; they know next to nothing of the real merits or demerits, the fitness or unfitness of the candidate for whom they vote, and—having voted—all further demonstrativeness is out of place and wrong.

And what has been said above respecting politics may be said equally of the Church and religion. All know how the Church, catholic as called and as it ought to be, is broken up into parties at the present day. To say nothing of foreign Churches, or of the many sectarian Churches not in harmony with the Established Church in this country, what differences of opinion exist among ourselves even in this last. High Church, Low Church, Broad Church—what do these names indicate but the extreme opposite views of the two first and the all-comprehensive views of the third? I give no opinion as to the particular views held by any of the three. I only mention them in order to show what parties there are existing in our own Church, as distinguished from all other Churches. And then comes the question, how are we to deal with such parties, how determine to which we shall ourselves belong? I do not see that we have any occasion to consider them at all in the light of parties, whatever we may think or have to say respecting their opinions. If you ask which Church you should belong to, I meet the question by referring to a passage in one of Paul's epistles, which, though relating to a very different matter, might well be applied to the matter before us:—"Brethren, let every man, wherein he is called, therein abide with God" (I. Cor. vii. 24). Let every man abide in the Church of his fathers—*i.e.*, in which he has been brought up and educated—till he sees good cause for leaving it. And so long as he abides in it, let him hold kindly intercourse with the men of all other Churches, so far as circumstances bring him and them together. Let there be no feelings of discord or anger, no show of uncharitableness, on one side or the other. They are the children of our heavenly Father, though their earthly fathers are different. They may discuss their respective views and opinions in a friendly way, but not in a party spirit. In such a case no man is so bound to a particular creed that he cannot get loose; only let his arguments rest on Scripture, or on what he believes to be a right interpretation of Scripture. The Apostle's advice in another passage is, "Prove all things; hold fast that which is good," only, "let all your deeds be done with charity."

It is sad to think of the contrast between these words and the spirit so often shown by writers in some of our papers and periodicals, each of our Church parties in turn vindicating its own opinions, and crying down (sometimes in the most intemperate language) those who differ from them.

This is not said in condemnation of all controversy. Controversy has its use, and, when conducted in a right spirit, it may lead in the end to a better understanding of things and the discovery of truth. "Whatever retards a spirit of inquiry," said Robert Hall, "is favourable to error; whatever promotes it, is favourable to truth. But nothing has greater tendency to obstruct the exercise of free inquiry than a spirit of party. There is in all sects and parties a constant fear of being eclipsed. It becomes a point of honour with the leaders of parties to defend and support their respective peculiarities to the last, and, as a natural sequence, to shut their ears against all the pleas by which they may be assailed. If we seek for the reason of the facility with which scientific improvements establish themselves in preference to religious, we shall find it in the absence of party combination."

RECOGNITION OF FRIENDS IN A FUTURE STATE.

It is the necessary consequence of living to a great age that a man loses nearly all his friends—at least, his old friends—before dying himself. As in my own case; there are but a very few left among my acquaintances of the same generation as myself. This circumstance leads one to reflect on the subject of the recognition of friends in a future state. There are many passages in Scripture which fully encourage us to look for such recognition. But I am not about to quote all these; my purpose is rather to consider the question in a metaphysical point of view. I would ground the answer on the fact of personal identity and on arguments deduced from such identity. I call personal identity a fact, for is it not a fact that we feel ourselves to be one and the same person all through life, whatever may have been its manifold changes? We remember ourselves as children, as youths, and as grown-up men and women, according to the age we have attained. We remember the chief events in each of those periods of life, and, taken collectively, this is personal identity. We know, moreover, that our condition in a future life depends upon the character of our present life; and we could not well be charged with having led an evil life or commended for a good and useful life, had we no recollection and were no recollection possible, as to how we had conducted ourselves here, ill or well, as the case may have been.

But further:—Along with other recollections of the past there must necessarily be classed the recollection of those among whom our life on earth was spent, friends and relatives from whom our life had received in great measure its tone and colouring. And if there were no recognition in a future state of those who had so largely contributed to our welfare here, to whom, in some cases, all that we possessed and all that we enjoyed in the way of support, comfort, and happiness was due, what of real happiness could we enjoy hereafter, under any conceivable conditions, with the thought that we were never again to have the companionship of those dear ones on earth, from whom we had so long been separated, whom we never could forget, but were never to see again?

I would here just refer to one passage of Scripture in connection with this subject. If we turn to St. Paul's Epistle to the Colossians (Col. i. 28), we find these remarkable words:—" Whom we preach, warning every man, and teaching every man in all wisdom, that we may present every man perfect in Christ Jesus." Paley, taking this passage as a text for one of his sermons, comments upon it as follows, by this, " I understand St. Paul to express his hope and prayer that at the general judgment of the world he might present them perfect in every good work. And if this be rightly interpreted, then it affords a manifest and necessary inference, that the saints in a future life will meet and be known again to one another; for how (he wisely remarks), without knowing again his converts in their next and glorified state, could St. Paul desire or expect to present them at the last day?"

This seems almost conclusive as to the question we

are considering. But we need not stop here. The argument from recollection may be extended to the inquiry suggested by Dean Plumptre, in his interesting work on "The Spirits in Prison and other Studies on the Life after Death." Speaking of the new body with which the saints shall be clothed, when the earthly house of this tabernacle is dissolved (II. Cor. v. 1—4), the Dean asks, "What are the conditions of that body, whether it is after the likeness of the man when he dies in infancy, or maturity, or age, or represents, so to speak, the ideal of his personal humanity?" These questions, he says, may be asked, but cannot easily be answered. Yet, as bearing upon the recognition of friends—I say nothing as to the exact nature of the spiritual body—it seems reasonable to think that their appearance to us would be in keeping with those same recollections upon which we ground our belief in the possibility of there being any recognition at all. Their appearance would be to us the same as when we last saw them. This, however, does not preclude the remembrance of them, as known to us at any earlier period of their lives, supposing we had known them a long time before they died. More than this, it would be rash to assert.

But again, in the same chapter of his book* above referred to, the Dean puts a second question of not less interest than the first:—"Do the souls of the righteous know what is passing on the earth? Do they think of

* Chap. xvi—"The Activities of the Intermediate State." See pp. 400 and 411.

and pray for those they have left behind them?" Of course, no certain answer can be made to such an inquiry any more than to the former one. But here, too, I think—without reference to Scripture—we may venture a conjecture, based, as before, on the argument from recollection. Assuming that our deceased friends have the same recollections of us that we have of them—it seems to follow that if they have no knowledge of us and our doings, there must be a constant longing after such knowledge, which, with the felt impossibility of obtaining it, must, one would think, diffuse a cloud, more or less, over whatever happiness, derived from other sources, may fall to their lot in their heavenly abodes.

It may be—speculations on this whole subject are of little worth—but they naturally engage the thoughts of an old man, long since parted from those most near and dear to him, soon however—now—"to enter into that which is within the veil," and realise whatever lies beyond it.

I conclude these Chapters with a List of the Scientific Societies to which I belong, and the year of Election.

Elected a member of the LINNEAN SOCIETY in Nov., 1822, and have been the *Father* of the Society for many years.

In the same year, 1822, elected a member of the CAMBRIDGE PHILOSOPHICAL SOCIETY.

I am an *original* member of the three following Societies:—ZOOLOGICAL, Instituted in 1826; ENTOMOLOGICAL SOCIETY, Instituted in 1834; and RAY SOCIETY, Instituted in 1844.

Joined the BRITISH ASSOCIATION for the Advancement of Science in 1832, being the second year of its existence.

Elected a member of the GEOLOGICAL SOCIETY of London in 1835.

In 1839, elected an *Honorary* Member of the BOSTON NATURAL HISTORY SOCIETY, U.S.

In 1841, elected an *Honorary* Member of the ROYAL ZOOLOGICAL SOCIETY OF IRELAND.

In 1849, elected an *Honorary* Member of the IPSWICH MUSEUM.

In 1855, I myself Instituted the BATH NATURAL HISTORY FIELD CLUB, and was chosen as President.

In 1863, elected a Member of the ANTHROPOLOGICAL INSTITUTE.

I have mentioned above* the chief works published by myself at different times. At the request of a friend I here supplement it with a complete List of my Papers in the Transactions of Scientific Bodies, Magazines, etc., including various Tracts, Notes, and short communications.

(1) Observations on the Ornithology of Cambridgeshire. (*Trans. Camb. Phil. Soc.*, vol. 2.)

(2) Observations on a Præternatural Growth of the Incisor Teeth, occasionally observed in certain of the Mammalia Rodentia. (*Loud. Mag. Nat. Hist.*, vol. 2, p. 134.)

(3) The Distinctive Character of two British Species of *Plecotus*, supposed to have been confounded under the name of "*Long-eared Bat.*" (*Linn. Trans.* vol. 16, p. 53.)

N.B.—The supposed new species proved to be only the young of *Plecotus Auritus*.

(4) Some Observations on the Habits and Characters of the *Natter-Jack* of Pennant, with a List of Reptiles found in Cambridgeshire. (*Trans. Camb. Phil. Soc.*, vol. 3, p. 373.)

(5) Some Observations on the *Common Bat* of Pennant, showing its identity with the *Pipistrelle* of French Authors. (*Linn. Trans.*, vol. 16, p. 159.)

(6) Some Remarks upon the *Winter* of 1829-30, and upon the general character of the weather which preceded and followed it. (*Loud. Mag.*, vol. 3, p. 538.)

(7) On a Peculiar Species of *Mite* parasitical on *Slugs*. (*Loud. Mag. Nat. Hist.*, vol. 4, p. 538.)

(8) An Extraordinary Swarm of *Flies*, Nov., 1831. (*Loud. Mag. Nat. Hist.*, vol. 5, p. 302.)

* See pp. 29—31.

(9) Monograph on the British Species of Cyclas and Pisidium. (*Trans. Camb. Phil. Soc.*, vol. 4, pp. 289-312.)

(10) Some Remarks on Genera and Subgenera, and on the Principles on which they should be established. (*Loud. Mag. Nat. Hist.*, vol. 6, p. 385.)

(11) On Designating Genera and Subgenera, and on the Principles of Classification which they involve. (*Loud. Mag. Nat. Hist.*, vol. 7, p. 97.)

(12) Systematic Catalogue of British Vertebrate Animals, 8vo. pp. 16, Camb. 1835.

(13) Report on the recent progress and present state of Zoology. (*Brit. Assoc. Report* for 1834, p. 143.)

(14) Some Remarks on the Study of Zoology, and on the present state of the Science. (*Jard. Mag. of Zool. and Bot.*, vol. 1, p. 1, 1836.)

(15) Catalogue of the Collection of British Quadrupeds and Birds, in the Museum of the Cambridge Philosophical Society, sm. 8vo, Camb. 1836.

(16) On the Dentition and other Characters of the British Shrews, with reference to M. Duvernoy's recent researches into the structure of this genus of animals. *Jard. Mag. of Zool. and Bot.*, vol. 2, p. 24, 1837.)

(17) Further Remarks on the British Shrews, including the distinguishing characters of two species previously confounded. (*Ann. Nat. Hist.*, vol. 1, p. 417, 1838.)

(18) Additional Notes on the British Shrews. (*Ann. Nat. Hist.* vol. 2, p. 43.)

(19) Notes on some Shrews brought from Germany by W. Ogilby, Esq., including the description of an apparently new species (*Ann. Nat. Hist.*, vol. 2, p. 323.)

(20) On a new species of Bat found in the County of Durham. *Ann. Nat. Hist.*, vol. 3, p. 73.)

(21) On three undescribed species of *Cimex*, closely allied to the common Bed-bug. *Ann. Nat. Hist.*, vol. 3, p. 241.

(22) Notice of the Museum of the Cambridge Philosophical Society. (*Cambridge Portfolio*, p. 127, 1840.)

(23) Biographical Notices of Willughby and Lister. (*Cambridge Portfolio*, p. 130.)

(24) Notice of a case in which the Larvæ of a Dipterous Insect, supposed to be the *Anthomyia canicularis*, Meig., were expelled in large quantities from the human Intestines; with description and figures of the same. (*Entomolog. Trans.* 1839, vol. 2, p. 152.)

N.B.—This case occurred in a patient of the late Dr. Haviland, of Cambridge, who sent me the particulars.

(25) Review of "Etudes de Micromamalogie ; par M. Edm. De Selys-Longchamps." (*Ann. Nat. Hist.*, vol. 4, p. 4-34.)

(26) Review of Yarrell's "British Birds." (*Lond. and Westmin. Rev.*, No. 65, March, 1840, p. 373.)

(27) Notes on some of the Smaller British Mammalia, including the description of a new species of *Arvicola* found in Scotland. (*Ann. and Mag. Nat. Hist.*, vol. 7, p. 261, Jun., 1841.)

(28) On the Reproduction of Frogs and Toads without the intermediate Stage of Tadpole. (*Ann. and Mag. Nat. Hist.*, vol. 11, 2nd ser., p. 482, 1853.)

(29) On the Turf of the Cambridgeshire Fens. (*Brit. Assoc. Report*, Cambridge, 1845, communications to the sections, p. 75.)

(30) On the Variation of Species. (*Brit. Assoc. Report*, Cheltenham, 1856, comm. to sections, p. 101.)

(31) Note on the Smaller British Species of *Pisidium*. (*Ann. and Mag. Nat. Hist.*, 3rd ser., vol. 2, p. 104, Aug., 1858.)

(32) Sketch of the Flora of Bath, 1864. (Contributed to "Wright's Historical Guide Book to Bath," pp. 401-415.)

(33) Address delivered to the Members of the Bath Natural History and Antiquarian Field Club, Feb. 2, 1864, 8vo, pp. 16.

(34) On the Temperature and Rainfall of Bath. (*Bath Chronicle Report of Proceed. of Brit. Assoc. Bath*, 1864, p. 108.)

(35) Natural History Museums. Lecture delivered to the Members of the Bath Nat. Hist. Field Club, Feb. 21, 1865, 8vo, pp. 24.

(36) Phosphatic Nodules, obtained in the Eastern Counties for Agricultural Purposes. Lecture to the Members of the Bath Field Club, April 12, 1866. 8vo, pp. 18.

(37) The Bath Flora. Lecture delivered to the Members of the Bath Field Club, Dec. 5, 1866, 8vo, pp. 39.

(38) Notes on the Summer of 1868, particularly the Temperature as observed in Bath, and compared with that of Greenwich and some other places. (*Proceed. Bath Nat. Hist. Field Club*, pt. 3, 1869.)

(39) Address at the Opening Meeting of the Bath Nat. Hist. Field Club, Jan. 12th, 1870. (From the "*Bath Express.*")

(40) St. Swithin and other Weather Saints. (*Proceed. Bath Nat. Hist. Field Club*, vol. 2, part 2, 1871.)

(41) Five Addresses to the Members of the Bath Nat. Hist. Field Club after the Anniversary Dinner, Feb. 18th, 1868, 1869, 1870, 1871, 1872. (*Proceedings of the Club.*)

(42) Local Biology; followed by Remarks on the Fauna of Bath and Somerset. (*Proceed. Bath Nat. Hist. Field Club*, vol. 2, part iv., 1873.)

(43) Copy of a Letter from Mr. Stephens of Camerton, near Bath, to Mr. Davis of Longleat, on the subject of the Diseases of Wheat, dated Aug. 22nd, 1800; with Remarks

by the Rev. L. Blomefield. (*Proceed. Bath Nat. His. Field Club*, vol. 3, p. 12, 1874.)

(44) Note on the Occurrence of the Land Planaria (*Planaria terrestris*) in the Neighbourhood of Bath. (*Proceed. Nat. Hist. Field Club*, vol. 3, p. 72, 1874.)

(45) Results of Meteorological Observations, made at the Bath Literary Institution during ten years, commencing with March, 1865, and ending with February, 1875. (*Proceed. Bath Nat. Hist. Field Club*, vol. 3, No. 3.)

(46) Gales of Wind. (*Proceed. Bath Nat. Hist. Field Club*, vol. 4, part 1, 1877.)

(47) The Winter of 1878-9 in Bath, and Seasons following. (*Proceed. Bath Nat. Hist. Field Club*, vol. 4, part 3, 1880.

(48) Case of Abnormal Development of Wood in the Root of a Spanish Chestnut. (*Proceed. Bath Nat. Hist. Field Club*, vol. 5, p. 21, 1882.)

(49) Note respecting the Winter of 1882-3. (*Proceed. Bath Nat. Hist. Field Club*, vol. 5, part 2, p. 111, 1883.)

(50) Notice of a Rare Capture, followed by Remarks on Variation of Structure and Instincts in Animals. (*Proceed. Bath Nat. Hist. Field Club*, vol. 5, part 3, p. 172, 1884.)

(51) Notice of a Second Capture of the Rare Longicorn, taken near Bath in September, 1883. (*Proceed. Bath Nat. Hist. Field Club*, vol. 5, part 4.)

N.B.—The two last papers have reference to a Rare Longicorn Insect taken in Bath, and thought at the time to be the *Cerambyx œdilis* of Linnæus, but which proved afterwards to be *Monochumus dentator* of Fabricius.

(52) The Bournemouth Firs; considered in connection with the probable existence of a Forest of Firs in that part of

England in Ancient Times. Was there ever a Forest of Firs on the Hills Around Bath? (*Proceed. Nat. Hist. Field Club*, vol. 6, p. 39, 1886.)

> *N.B.*—The above Title perhaps would be more correct than the one affixed to the paper originally, as a large number of the Firs now existing at Bournemouth were planted in the early part of the present century.

(53) Address to the Members of the Bath Field Club in reference to the death of C. E. Broome, Esq,, F.L.S., an Original Member of the Club. (*Proceed. Nat. Hist. Field Club*, vol. 6, p. 144, 1886.)

(54) Letter to the Editor of the *Bath Chronicle*, July 7, 1887, on the "Selborne Society."

(55) Further Results of Meteorological Observations made at the Bath Royal Literary and Scientific Institution. (*Proceed. Bath Nat. Hist. Field Club*, vol. 6, p. 185, 1887.)

> *N.B.*—The above is in continuation of a former paper (see No. 45). The two papers together give the results of Meteorological Observations carried on for 20 consecutive years, without break, at the Bath Literary Institution.

(56) Reminiscences of William Yarrell, sm. 8vo, pp. 12, Bath, 1885.

(57) Reminiscences of Prideaux John Selby, and Twizell House; also Brief Notices of other North Country Naturalists, sm. 8vo, pp. 39, Bath, 1885.

(58) Chapters in My Life, 8vo, pp. 74, Bath, 1887.

> *N.B.*—The above three last Tracts were privately printed.

For EU product safety concerns, contact us at Calle de José Abascal, 56–1º,
28003 Madrid, Spain or eugpsr@cambridge.org.

www.ingramcontent.com/pod-product-compliance
Ingram Content Group UK Ltd.
Pitfield, Milton Keynes, MK11 3LW, UK
UKHW040158230326
469255UK00012B/160